Evolutionary Economics and Environmental Policy

NEW HORIZONS IN INSTITUTIONAL AND EVOLUTIONARY
ECONOMICS

Series Editor: Geoffrey M. Hodgson
Research Professor, University of Hertfordshire Business School, UK

Economics today is at a crossroads. New ideas and approaches are challenging the
largely static and equilibrium-oriented models that used to dominate mainstream
economics. The study of economic institutions – long neglected in the economics
textbooks – has returned to the forefront of theoretical and empirical investigation.

This challenging and interdisciplinary series publishes leading works at the forefront of institutional and evolutionary theory and focuses on cutting-edge analyses of
modern socio-economic systems. The aim is to understand both the institutional
structures of modern economies and the processes of economic evolution and development. Contributions will be from all forms of evolutionary and institutional economics, as well as from Post-Keynesian, Austrian and other schools. The overriding
aim is to understand the processes of institutional transformation and economic
change.

Titles in the series include:

Evolutionary Economics and Environmental Policy

Survival of the Greenest

Jeroen C.J.M. van den Bergh

Faculty of Economics and Business Administration, and Institute for Environmental Studies, Free University, Amsterdam

Albert Faber

Netherlands Environmental Assessment Agency, Bilthoven

Annemarth M. Idenburg

DHV Group, Amersfoort, formerly, Netherlands Environmental Assessment Agency, Bilthoven

Frans H. Oosterhuis

Institute for Environmental Studies, Free University, Amsterdam

NEW HORIZONS IN INSTITUTIONAL AND EVOLUTIONARY ECONOMICS

Edward Elgar
Cheltenham, UK • Northampton, MA, USA

Published by
Edward Elgar Publishing Limited
Glensanda House
Montpellier Parade
Cheltenham
Glos GL50 1UA
UK

Edward Elgar Publishing, Inc.
William Pratt House
9 Dewey Court
Northampton
Massachusetts 01060
USA

A catalogue record for this book
is available from the British Library

Library of Congress Cataloguing in Publication Data

Evolutionary economics and environmental policy : survival of the greenest /
Jeroen C.J.M. van den Bergh — [et al.].
    p. cm. — (New horizons in institutional and evolutionary economics)
    "This book is the result of a joint effort of the Netherlands
Environmental Assessment Agency, the Faculty of Economics and Business
Administration of the Free University (Vrije Universiteit) in Amsterdam, and
the university's Institute for Environmental Studies."
    Includes bibliographical references and index.
    1. Energy policy—Netherlands—Case studies. 2. Energy policy—
Environmental aspects. 3. Evolutionary economics. 4. Environmental
policy—Netherlands. 5. Sustainable development—Netherlands. I. Bergh,
Jeroen C.J.M. van den, 1965–

HD9502.N42E89 2007
333.701—dc22   2006022405

ISBN 978 1 84542 955 3

Printed and bound in Great Britain by MPG Books Ltd, Bodmin, Cornwall

Contents

Figures

Tables

Boxes

Preface

This book is the result of a joint effort of the Netherlands Environmental Assessment Agency, the Faculty of Economics and Business Administration of the Free University (Vrije Universiteit) in Amsterdam, and the university's Institute for Environmental Studies. This has led to a thorough search for the potential contribution of evolutionary economics to understanding innovations and transitions in the context of energy and sustainable development. This in turn has given rise to practical suggestions as to the role of the government and the design of public policies aimed at making a transition to a sustainable development. Current policies in a number of relevant areas are critically examined against the background of lessons learned from the application of evolutionary economics. In addition, three specific energy technologies – namely, fuel cells, nuclear fusion and photovoltaic energy – are examined in detail within the adopted evolutionary-economic framework. A glossary of the evolutionary-economic terminology employed here is included at the end of the book.

We are very grateful to many colleagues for constructive comments on earlier drafts of the text: René Kemp (MERIT, Maastricht), Frank Dietz (Ministry of Housing, Spatial Planning and the Environment), Marlouke Durville (Ministry of Economic Affairs), Marjan Hofkes (Institute for Environmental Studies) and Rob Maas, Fred Langeweg, Leon Janssen, Jan Ros, Harry Wilting, Eric Drissen and Jacco Farla (all Netherlands Environmental Assessment Agency). Frans Kooymans translated the text from Dutch into English.

It is our hope that this book will serve as a stimulus for future research on the implications of evolutionary economic thought for environmental policy and transition management.

Jeroen van den Bergh
Albert Faber
Annemarth Idenburg
Frans Oosterhuis

1. Introduction

Economic premises and principles play a major role in the shaping of environmental policy. In traditional neoclassical economics, concepts such as rationality, efficiency and optimization are dominant. This leads to an approach that turns out to be unsuitable for the analysis and interpretation of major system changes (transitions) and innovations. Other economic theories provide complementary but sometimes contradictory viewpoints. The most valuable theoretical framework for the study of innovations and transitions is *evolutionary economics*, with its key concepts of diversity, innovation and selection. The starting premise of evolutionary economics is that human beings act according to bounded rationality, which takes the form of routines, imitation and a limited time horizon. Especially in studies of technological development and innovations, evolutionary economics plays an increasingly important role. In this study we are especially interested in the common grounds of environmental and innovation policy. The key question of this book can be formulated as:

> *What insights does evolutionary economics theory provide for the design of environmental policy that aims to stimulate innovations and a transition to a long-term environmentally sustainable economy?*

In this study we will focus our attention on energy; in particular, on policies for energy innovation and an 'energy transition' to a sustainable energy provision. The key question is therefore translated into three subordinate questions:

1. What concepts and ideas from evolutionary economics can be applied to develop a vision on environmental policy and transition management, how can these concepts be applied and which results can thus be derived?
2. To what extent are the views on policy with regard to stimulation of energy innovations – as reflected in today's policy documents and advice – in line with the insights of evolutionary economics?
3. How can the development of specific energy technologies be understood from an evolutionary-economic perspective?

The organization of the remainder of this book is as follows. Chapter 2 provides a general, theoretical overview of the principles and main insights of evolutionary economics. This culminates in a choice of key concepts and a synthetic framework for evolutionary-economic analysis. In Chapter 3 this framework is used to derive a number of suggestions for the design of environmental policy and transition management. Next, in Chapter 4 an evaluation of current energy and innovation policy in the Netherlands is carried out on the basis of the evolutionary-economic principles derived in the initial chapters. In Chapter 5 the general insights are illustrated and elaborated within three case studies that are relevant to the transition to a sustainable energy supply. Three technologies are looked at in detail; namely, fuel cells, nuclear fusion and photovoltaic energy. Chapter 6 summarizes and draws main conclusions from this study.

2. Evolutionary economics

2.1 EVOLUTIONARY THINKING

Evolution is nowadays widely regarded as both a general concept and a set of concrete mechanisms to comprehend structural change processes that affect human technologies, organizations and institutions (e.g. Ayres, 1994; Dennett, 1995). The application of evolutionary thinking to economics has made significant headway during the past 20 years. This is partly the result of dissatisfaction with the way technological development is presented in the neoclassical economic model of economic growth. Technology was originally viewed as an exogenous variable that develops outside the economic process. Later it came to be viewed as an endogenous variable that could be fitted rather easily into the framework of economic equilibrium analysis. More recent insights into the process of innovation, however, undermine this latter assumption. Technological development is now viewed as the outcome of a continuous interaction between generation (innovation) and selection of diversity in technologies and organizational structures. When we take this view as a starting point, the evolutionary approach provides a credible alternative to traditional theories of economic and technological change. The rational behaviour of individual persons and groups, which is assumed in traditional economic theory, is replaced in the evolutionary approach by bounded rationality, which can take the form of habits, routines, myopia and imitation (Nelson and Winter, 1982; Robson, 2001).

The following elements and processes play key and complementary roles within evolutionary economics:

- *Diversity* (variation): populations of strategies, products, technologies and organizational structures.
- *Selection*: processes that reduce existing diversity.
- *Innovation*: processes that create more (increase) diversity.
- *Transferability (transmission)*: replication by reproduction or copying (possibly imitation). This results in continuity, durability (retention) and cumulative processes.
- *Bounded rationality*: individuals and organizations (groups) largely behave according to fixed patterns that result from adjustments in the

3

past to a certain environment or that have been selected by this environment. Characteristic features of this are routine behaviour, imitation of others and myopia (a short time horizon).

Paying specific attention to diversity implies an approach that takes agent populations as the starting point. The representative agent, which is a customary assumption in neoclassical economic theory, thus no longer applies. A population approach can be translated into a formal model. The type of model depends heavily on the scope of aggregation. Evolutionary game theory, for example, depends on aggregate variables, so that the extent of diversity is generally confined to two or three characteristics. An alternative approach describes populations and related changes by means of probabilistic distributions. A third mode, the ultimate micro approach, describes individuals on the basis of unique characteristics. This results in the most detailed description of possible interactions between economic agents, also known as multi-agent systems. These can then be placed in a context of fully random interactions (a gas cloud) or of systematic interactions in a network or spatial (e.g. cellular grid) structure.

Despite the apparent simplicity of the underlying mechanisms, evolution is a powerful theory. In essence it consists of two diametrical forces or causal processes. One involves the creation of diversity through the application of various mechanisms, often combined under the terms discovery and innovation. This can be viewed as a force that stimulates imbalance. The second force is selection, which leads to a reduction of diversity. This can be regarded as a focused force that stimulates balance. Unlike a physical law such as the force of gravity, selection represents a taxonomy of a large number of processes that affect the fitness of individual elements, through their survival, reproduction and diffusion. In other words, it is an umbrella term. In an economic context it encompasses market competition, interactions between employers and employees (or trades unions), relationships with other pressure groups, mergers and acquisitions, financial requirements imposed by providers of capital (shareholders and banks), legislation and public regulation, and public opinion.

The result of these opposing forces of innovation and selection is a process of continuous change, without this necessarily leading to a state of equilibrium. Only with limited innovation or no innovation at all can a system converge under the influence of selection processes to a state of balance. This is the approach followed by evolutionary game theory. Diversity will then necessarily decrease until it reaches a minimum level. In a state of balance between selective and innovative forces, diversity will sometimes increase and at other times decrease, but will never be fully eliminated. In addition, the dynamics of evolution depends on the existing diversity. The reason is that the outcomes of both innovation and selection processes depend on, or are limited by, the current state of diversity. In the case of innovation this is clear from the fact that many innovations are the result of combining already existing elements from a pool characterized by internal diversity ('recombination'). From the foregoing it is evident that neglecting to describe diversity in an evolutionary system would lead to incomplete comprehension of the dynamics of such a system.

An important consequence of evolution over an extended period of time is that structure and complexity come about. Evolutionary theory thus explains how processes at a single level generate new structures at a higher level. This is sometimes identified by terms, not altogether sharply defined, such as self-organization and emergence (Kauffman, 1993; Holland, 1998). Evolutionary theory provides an important overall basis for explanations of these phenomena. The success of evolutionary theory is evident not only within biology, its traditional area of application. The field of evolutionary computation within computer and information sciences, which solves design and optimization problems through the application of evolutionary algorithms, is an example of the effectiveness and practical usefulness of evolutionary theory and modelling (Bäck, 1996).

If an evolutionary system involves a great deal of internal diversity – in other words, a large population with many different elements or individuals – it is very unlikely, on the basis of mere probability theory, that the system will return to a previous state. We can then in fact speak of a situation of irreversibility, which implies that history is introduced. A major feature of the evolutionary approach is indeed that it merges theoretical and historical perspectives, notably since it incorporates causal mechanisms such as innovation, selection and accumulation. The theoretical consequences of this are path dependence and lock-in, which are described in detail later in this book.

The power and appeal of an evolutionary approach is that, despite its rather simple conceptual starting point, complex structures can be understood and explained on the basis of partially endogenous processes (innovation, and to a lesser extent, selection) that convert simple systems into complex ones. It can therefore safely be asserted that evolution is one of the most powerful and comprehensive ideas that science has brought about (Ayres, 1994; Dennett, 1995). As evidence of this, evolutionary theory has been extensively applied within modern biology for more than half a century and with tremendous success. For several decades there has also been a growing acknowledgement of the potential of evolutionary thinking within the social sciences. A fascinating application of evolutionary (multi-agent) modelling, for example, is the by now famous Sugarscape model of Epstein and Axtell (1996), which describes the development of a complex socio-economic system, combining elements from economics, demography, sociology and biology. Comparable evolutionary models have meanwhile also been successfully applied to the financial markets (Levy et al., 2000). This book, which focuses on the analysis of environmental and energy policy, is in line with these developments.

2.2   A BRIEF OVERVIEW OF IDEAS AND CONCEPTS WITHIN EVOLUTIONARY ECONOMICS

Evolutionary economics builds upon the general evolutionary principles that have been outlined in the previous section. The development of this field features a great variety of contributions that have led to new concepts and ideas that may be useful in research on environmental and transition policy. This section presents a brief overview of the main contributions and insights within evolutionary economics. A complete and detailed overview can be found in Hodgson (1993). An accessible introduction in Dutch is presented in Boschma et al. (2002).

2.2.1   Origins

Veblen (1898) is often regarded as the first evolutionary economist, based mainly on the fact that he used the term 'evolution' explicitly in a discussion of how to approach the study of economics. In particular, he considered the question of why economics had not developed into an evolutionary science. Veblen's approach was very sociological, as he emphasized the behaviour of entrepreneurs. He paid much attention to technological change, the pace of which was rapid around 1900, but which altogether lacked a coherent theory. His suggestion of evolutionary economics was one of a causal process, an unfolding of phases, an accumulation of effects.

Joseph Schumpeter was the most influential of all early evolutionary economists, both because of his reputation within the European and American economic communities and because of the many concepts and ideas that sprang from his mind. Schumpeter questioned the dominance of the static approach to economic science. This linked up with a considerable interest in the dynamics of economics; in particular, the development of the capitalist system (as also studied by Karl Marx), which was evident in all of his important works (Schumpeter, 1934, 1939, 1942). In his book *The Theory of Economic Development* (Schumpeter, 1934; originally published in German in 1911) he viewed qualitative economic and technological change within the broader context of social change, with an emphasis on psychological aspects relating to the influence of innovative entrepreneurship. Schumpeter believed in the value and the ideal of equilibrium analysis, which was somewhat inconsistent with his search for a dynamic evolutionary theory. A possible explanation for this is that his views changed during the course of time.

Schumpeter regarded economic (meaning capitalist) change as the result of revolutionary forces within the economy that destroy old processes and create new ones. He referred to this as *creative destruction*, a term often quoted later. This process enables discrete or non-gradual changes, far removed from any state of equilibrium. These are reinforced by clusters of innovations that succeed and are derived from a major innovation. Schumpeter developed these themes further in his studies of business or economic cycles; in particular, of what he referred to as *long waves*. Schumpeter shares with the classic economists Marx, Mill and Ricardo the generic idea of a steady state of stable, gradual changes that the economy will ultimately end up in. In Schumpeter's case this is characterized by technical progress that results from carefully planned team research within a socialist organization of the community at large.

Another major concept that arose from Schumpeter's dynamic perspective is what later came to be referred to as Schumpeterian competition, as opposed to equilibrium or price competition. This refers to the achievement of a competitive advantage by a business enterprise over its competitors through early innovation or adoption of a new product or process (technology). Later, this concept came to be integrated into the Porter hypothesis, which can be regarded as a type of Schumpeterian policy competition among countries. In particular, countries can achieve competitive advantage through innovation under stringent environmental regulation (Porter and van der Linde, 1995).

Even though Schumpeter's writings do not explicitly refer to evolutionary concepts and terminology, his work is an important starting point for modern evolutionary economics. Nonetheless, Schumpeter did not altogether avoid the term 'evolution': ' . . . the essential point to grasp is that in dealing with capitalism we are dealing with an evolutionary process' (Schumpeter, 1954, p. 82; quoted in Potts, 2000). It is fair to say that both Veblen and Schumpeter were strongly influenced by the changes that took place around them in terms of economic structure and technical innovations. After all, these changes involved a great deal of diversity and a high rate of innovation.

2.2.2 Profit Maximization through Market Selection?

Since 1950 there has been a steady increase in the number of publications about economic evolution. This might be explained by the success of evolutionary biology, but also by the limitations and growing criticism of the assumptions that neoclassical economics runs up against when facing questions such as how to deal with technological change. In addition, there is the search for the evolutionary basis of optimizing behaviour that is assumed in neoclassical economics. An early attempt to answer this can be found in the work of Alchian (1950) and the 'translation' thereof by Friedman (1953). Alchian reasoned that it was unnecessary to view profit maximization or even the search for profit as a fundamental and universal characteristic of enterprises, since companies that seek profit and are successful in this respect will be selected (Alchian uses the term 'adopted') by the working of the market: that is, only they survive. Whether companies are successful is mainly decided by chance, regardless of whether or not they search for profit, and the less certain the world, the more profits will depend on pure chance and not on the calculated search for profit. Friedman (1953) went beyond Alchian by positing that markets select effective or successful profit maximization behaviour and not simply 'seeking and generating profits'.

Winter (1964) criticized Alchian and Friedman for their use of the concept of selection, arguing that they failed to consider transmission mechanisms that determine that successful behaviour can be maintained and copied over time. Without such mechanisms it is impossible for markets to select companies that consistently succeed in achieving profits, let alone maximization of profits. Without transferability of specific entrepreneurial behaviour, profit in a specific period will be unrelated to profit in another period. If making or maximizing profit is not the result of an explicit and conscious set of procedures, then it cannot be transferred to others or learned by them. As a consequence. making a profit is largely the outcome of chance, being dependent on many uncontrollable factors. This view is illustrated by the fact that the profit level of many companies shows irregular fluctuations over time (often even turning into a loss).

A further consideration is that when a transfer of behaviour takes place, uncertainty and changes in the economic environment act as a disrupting factor in the selection process. In other words, selection does not always lead to the best or to the survival of the fittest. The result is that the market will fail to select profit makers or maximizers in a consistent way.

Lastly, the mechanism of market selection itself is far from perfect. In the case of weak competition, for example, selective pressure will be limited. Also, there are generally many additional selective forces that reduce the impact of market forces. Examples are the many types of legislation and government regulations that are not intended to select profit generation (Foss, 1993).

Contrary to Alchian, Friedman (1953) had an ambitious objective; namely, to identify a clear methodological approach for economic science.

In a certain sense he was successful at this, and not just because his essay is among those frequently quoted in the field of economic methodology. In particular, evolutionary game theory, which arose during the 1980s, links up well with the ideas of Alchian and Friedman (see Section 2.2.5). Both emphasize equilibrium selection of existing diversity and disregard structural processes of innovation that cause the system to get out of balance. This generally leads to the conclusion that a system ends up in a state of equilibrium, which according to Friedman's line of thinking consists entirely of profit maximizers.

2.2.3   Routines and Search Behaviour: Nelson and Winter

The most quoted and possibly also most influential work within the field of evolutionary economics since the 1950s has been that by Richard Nelson and Sydney Winter, which culminated in their famous book published in 1982, *An Evolutionary Theory of Economic Change*. It has not only influenced evolutionary economists in the neo-Schumpeterian tradition but also gained the recognition of mainstream economists. An important reason for this lies in the combined theoretical–empirical nature of their work, grounded in a formal axiomatic approach.

Nelson and Winter express as their starting point: '. . . a major reconstruction of the theoretical foundations of our discipline is a precondition for significant growth in our understanding of economic change'. They focus their attention on the routines of companies and on gradual change. Their assumption about the core of companies is: '. . . motivated by profit and engaged in a search for ways to improve their profits, but their actions will not be assumed to be profit maximising over well-defined and exogenously given choice sets'. As to their analysis, they state: '. . . we do not focus our analysis on hypothetical states of "industry equilibrium", in which all the unprofitable firms are no longer in the industry and the profitable ones are at their desired size' (Nelson and Winter, 1982, p. 4).

The approach taken by Nelson and Winter can be summarized as gradual economic evolution based on changes in routine behaviour at a micro level that result from search processes. The three core elements of their theory of micro-evolution are organizational routines, search behaviour and selection environment. The first two elements are covered below, while the selection environment will be addressed in Section 2.4.4.

Organizational routines
Organizational routines refer to the way in which companies function and make decisions. A routine can be regarded as equivalent to the gene in biological evolution. It is a complex synthesis of the skills of employees and of

their mutual relationships, such as communication, collaboration and joint agreements. Skills or expertise can be defined as acquired and tacit knowledge, and feature a certain level of automation. Both skills and routines can be regarded as programmable. The difference between the two is that skills can be viewed as an individual feature, whereas routines apply to the level of organizations as a whole. A routine thus refers to the combination of individuals with specific and unique skills as well as to their mutual interactions. It thus describes a complex collection of interactive skills.

The interactions within a routine are crucial and depend on earlier contacts (learning, adjustment) and on 'language' specific to the organization. Similar to a skill, a routine is 'internalized' by carrying it out: this might be called 'remembering by doing'. The memory of an organization thus cannot simply be reduced to the sum of individual memories, since this would exclude interactions between professionals. Organizational memory can also not be reduced to blueprints, since that would ignore both the tacit nature of individual skills and the informal character of many (crucial) interactions between business employees. According to Nelson and Winter (p. 104), such blueprints may, however, be useful: '. . . perhaps, as a checkpoint to assess what might be wrong when the routine breaks down'. Not a single individual, including the 'boss', can possess all the information that is needed to allow the organization to function. The totality of expertise and past experiences is neither on paper nor in a computer file, and the 'boss' is not even aware of all the details involved. According to this view, an organizational routine operates as a distributed and self-organized system, without anyone in the organization needing to be aware of it, and without even the need for a hierarchical organizational structure. Evidently, with more individuals or more organizational complexity, a given task can be divided between more human individuals (Nelson and Winter, 1982, p. 106). This means that all or some individuals have a simpler task, or that tasks can be performed more accurately.

Change in routines through search behaviour

Routines create consistency and continuity in the behaviour of companies. Routines are in fact quite durable and stable for a number of reasons, amongst which are politics, avoidance of conflicts, vested interests, financial transaction costs of change and hierarchical management structure (Nelson and Winter, 1982). Nonetheless routines can change. This is the second element of Nelson and Winter's framework. Change of routines can take place in a number of ways. An evaluation of existing routines leads to small or large changes in routines or even to downright replacement. Organized search processes based on individual organizational units (which can themselves also function according to specific routines) form an important activity aimed at the change of existing routines. This potentially encompasses everything from products and processes to marketing strategies, internal organization and so on. Change of routines means that the behaviour of companies over time is characterized by flexibility and adjustment. It should be noted that 'routine innovation' of this type can itself also be subject to innovation. This was a major development after the Industrial Revolution, as entrepreneurs came gradually to be replaced by large corporations with R&D departments. This has affected the rate and direction of effective search and thus the rate of routine change.

Many changes in routines are, however, undirected and incidental. They may be the result, for example, of solving concrete problems in the organization or in its performance, or simply because employees leave and are replaced by others who possess different skills and create new interaction patterns with colleagues. An important change category is formed by new combinations of existing routines (Schumpeter, 1934) or cross-fertilization, which is analogous to the notion of 'recombination' in biological evolution. According to Nelson and Winter (1982, p. 130) this can take the form of '. . . new patterns of information and material flows among existing sub-routines'. This is achieved most effectively by taking routines that have proved reliable as a starting point, so that experiments via trial and error can focus on the identification of problems associated with the new combination rather than on errors in existing component routines. This results in a cumulative process of improvement via a combination of selection and innovation.

Another pattern that explains many changes in routines is replication or imitation of products, processes, strategies or organizational structures of other companies. Perfect replication is rare, since information about the routine to be copied is generally incomplete. Nelson and Winter argue that replication will therefore inevitably involve mutations. The complexity of a routine can easily be underestimated – given the amount of technology, expertise, number of people and communication patterns involved. In view of mutations during the replication process, Nelson and Winter recommend regarding the imitator as an innovator, since imitation and innovation are combined.

The framework of Nelson and Winter uses bounded rationality and automatic behaviour as a general business model, thus arguing that the neoclassical economic model of conscious and rationally considered choice between clearly defined alternatives does not correspond to real life. Companies can certainly operate within a certain interval of flexibility, but this is the result of routines rather than of consistent optimization under known conditions (Nelson and Winter, 1982, p. 126).

2.2.4 Current Schools of Thought (1): Neo-Schumpeterian Theories

Neo-Schumpeterian theories of technological change currently dominate evolutionary economics (Dosi et al., 1988; Witt, 1993; Metcalfe, 1998). They study phenomena at the company level (technological innovation), the market and sector level (competition and diffusion, structural change) and the macro level (growth, long waves and international trade). The impact of innovations at the company level is manifold. Innovations cause asymmetry between companies, sectors and countries in available technology, thereby enabling trade and commerce. Comparative advantages are not permanent but change as a result of innovation and diffusion. Trade in turn contributes to the spread of knowledge and technology. Technological change further-more impacts the division of labour, the organization of relationships within and between companies, and thus the industrial structure and patterns of intermediary supplies. This goes beyond structural changes that are studied using applied general equilibrium models. In such models the level of diversity (within sectors and technologies) is fixed, whereas within the evolutionary approach processes of innovation and selection give a more dynamic level of diversity. Interactions between consumers and producers, such as geographical and cultural proximity, are relevant as they allow the emergence of national or regional systems of innovation. Furthermore, some companies aim to expand their scope of activities and products, not only to realize economies of scope, but also to be resilient against market competition that leads to the destruction of outdated products.

The neo-Schumpeterian school of thought, stimulated by the work of Nelson and Winter, has resulted in a flood of literature about technological change. Key concepts include invention, innovation, adoption (imitation) and diffusion (sectoral or spatial). It can roughly be said that an innovation is always preceded by an invention that involves a character of surprise. Innovation is a cumulative process that involves uncertainty. Innovations can be categorized as follows:

- *Incremental*: continuous, learning-by-doing or using, efficiency improvements.
- *Radical*: discontinuous, combines product, process and organizational innovations.
- *Technology system changes*: various sectors impacted in a significant way.
- *Techno-economic paradigm changes* (technological revolutions): impact on the whole economy.

2.2.5   Current Schools of Thought (2): Evolutionary Game Theory

A second school, which is starting to gain more adherents, is evolutionary game theory (Weibull, 1995; Samuelson, 1997; Friedman, 1998a,b). This approach bears a relation to three preceding evolutionary methods. The first one is the work, already discussed, of Alchian and Friedman, who attempted to found equilibrium theory on an evolutionary or, better, selectionist perspective. Evolutionary game theory is also known as equilibrium selection theory, referring to the problem that non-linear models often result in a number of Nash equilibria. Adding evolutionary dynamics is thus a way to reduce the number of (feasible) equilibria. Next, there is a relationship with a group of well-known Chicago economists, who in the 1970s studied evolution and in particular selection, picking up ideas from sociobiology, as they developed a theory of utilitarian altruism (Becker, 1976; Hirshleifer, 1977; Tullock, 1979). The third approach is the method of evolutionary game theory as developed within biology (Maynard Smith, 1964, 1982; Maynard Smith and Price, 1973). This method was originally used theoretically to support insights of sociobiology.

Evolutionary game theory stresses the existence of asymptotic equilibria. These are enabled because of the absence of a structural innovation process, causing selection to fully dominate evolution. Diversity is thus reduced until a state of equilibrium with minimal diversity is achieved. The following statement by Foster and Metcalfe (2001, p. 9) is relevant in this regard: 'One intriguing aspect of evolution is that it consumes its own fuel. Processes of competitive selection necessarily destroy (or rather absorb) the very variety on which evolution depends. Unless variety is replenished, evolution will come to an end.' The focus of evolutionary game theory on selection also enables formal analytical solutions of equilibria.

Various equilibrium concepts are discussed in the literature on evolutionary games. Evolutionary equilibrium indicates that the population distribution of relevant characteristics does not alter, meaning that all characteristics have equal fitness. Economic (Nash) equilibria mean that economic agents, given the choices of others, cannot raise their profit or utility by altering their choices. The economic equilibrium is determined by applying a static model. Evolutionary equilibrium, on the other hand, is determined dynamically, since it is the result of a dynamic path that follows an initial population with a certain composition of characteristics (diversity). Such dynamics often occur on the basis of what is called replicator dynamics, in which individuals with above-average fitness see their share in the population rise, while those with below-average fitness see their share drop. Relative fitness is thus crucial and dependent on the diffusion of characteristics in the population. A key concept in evolutionary game theory is 'evolutionary stable strategy'. This combines characteristics of a Nash equilibrium with stability requirements. Stability is then defined as structural effects failing to occur in response to a single (exogenous) disturbance of a state due to a small change in the population distribution.

2.2.6   Connections, Hyperstructures and Discrete Mathematics

A recent proposal for the direction that evolutionary economics might take comes from Potts (2000). He established a link between evolutionary economics and network theory concepts. In his view, economic systems are complex hyperstructures or nested sets of connections between separate components. This may apply just as well to relationships between components of technology (cars, computers, factory machines) as to material and communication flows between individuals or departments in an organization (such as a company). Nelson and Winter's notion of business routines as interactions between individuals with specific skills also ties in well with the hyperstructures of Potts. Against this background, economic change and growth of knowledge are essentially processes of change in connections. As an example, computer chips are being applied everywhere, thereby creating new connections with already existing technology, such as television, audio equipment, cars, washing machines or telephones. As a result, this equipment functions more effectively and sometimes even takes on new tasks. The Internet, with its totally unique hyperstructure, forms a very concrete example of the added value of connections between people and (problem solving) knowledge.

In line with this idea of changing connections, Potts advocates a new micro-economics based on the technique of discrete and combinatorial mathematics, such as network or graph theory. This would cover the already available multi-agent evolutionary models, also known as 'artificial life' or 'artificial world' simulation models (not to be confused with traditional multi-sector models). Potts further asserts that many connections have a spatial dimension. He regards traditional micro-economic equilibrium theory as an approach that assumes a continuous reality, which is handy since it allows the application of concepts such as equilibrium, production function and utility function, plus techniques such as integration and differentiation. He argues, however, that this approach cannot properly deal with concepts such as complexity, heterogeneity, modularity and decomposition, nor with changes therein.

The network approach of Potts, on the other hand, is capable of handling complexity due to its dynamic interpretation of the balance between order and chaos. This concept follows the work of Kauffman (1993). Order corresponds here with the presence of a few connections between system components, while chaos is evidence of the presence of numerous connections. Complexity appears in a certain range of the number of connections between the two extremes. 'Underconnection' can be caused, for example, by path dependence and means an inflexible, non-adaptive system with isolated, independent elements. 'Overconnection', such as exists in unregulated financial markets, means continuous and unpredictable change (chaos), since very many or even all components react directly to each other (for example, because many people act upon the same information received via the media). Complexity, which lies in between, is evidence of a system with a relatively stable structure that has the capacity – within certain boundaries – to adjust to both external and internal changes.

In neoclassical economics, with its mathematical approach characterized by continuity, integration and differentiation, everything relates to everything, with the result that no distinction can be made between simple systems (with few connections) and complex systems (with many connections). Similarly, recognition of formation of structure (grouping, emergence) is impossible within such a context. Instead, neoclassical economic theory (and its models) assumes that the economic system possesses a fixed structure with only one level of interactions – thus there is no distinction between individuals and groups, and no emergence of new levels.

2.3   EVOLUTIONARY GROWTH THEORY

2.3.1   General Characteristics and Models

A major element of evolutionary economics, relevant to the search for sustainable development, is evolutionary growth theory. This section sets out the assumptions and insights of evolutionary views on economic growth analysis. In addition, it identifies the main differences between evolutionary and endogenous (neoclassical) economic growth theories.

The essence of an evolutionary theory of economic growth is that it has a bottom-up structure, as expressed by a description of a population of heterogeneous companies. This leads to differential growth, which can be regarded as a change in the frequencies of all potential individual properties. Growth thus inevitably goes along with a change in the mutual composition of business activity, just as in the real world. Nelson and Winter (1982, part IV; in particular, chapter 9) developed the first formal evolutionary model of economic growth, which can be compared with the famous Solow growth model dating from 1957. The model created by Nelson and Winter was intended to generate patterns of aggregate outputs, inputs and factor prices. Changes in the condition of a sector follow probability rules, with probabilities depending on search behaviour (aimed at improvements in profit or other performance indicators), imitation, investments, sector entry and selection. Only when companies make sufficient profit will they refrain from searching or imitating. Search behaviour is local, which implies minor improvements and staying close to current technology. Imitation can be directed at the average or at 'best practices'. This evolutionary growth model is a culmination of Nelson and Winter's theory of companies that act on the basis of routines and search behaviour and are subject to selection pressures. The result is a growth theory with an explicit link to evolutionary micro theory.

A key concept within neoclassical economic growth theory is the aggregate production function. This denotes a relationship between inputs and outputs of economic production activities at a very aggregate macro level; for example, aggregate income as a function of total labour and capital in a country. Nelson and Winter argue that ' . . . movements along the production function into previously inexperienced regions – the conceptual core of the neoclassical economic explanation of growth – must be rejected as a theoretical concept'. A common objection against continuous and distinguishable production functions is that neither individual companies nor entire sectors can move along the entire function, since information or knowledge only covers a limited number of production techniques. Evolutionary growth theory recommends avoiding an aggregated production function by maintaining the diversity of production relationships at the level of individual companies. Because of the resulting model complexity, this usually calls for a numerical analysis approach.

Nelson and Winter criticize neoclassical economic growth theory because it explains only some 20 per cent of productivity growth, based on movements along an aggregated production function resulting from factor input changes. The remaining 80 per cent, which thus remains unexplained, is often referred to by the term 'total factor productivity', and includes technological changes and general changes in environmental and resource factors. Nelson and Winter's approach allows for integration of the micro and macro aspects of technology and technological change. The results are not only consistent with the decisions of companies (routines, search behaviour) but also with empirical observations, as indicated by aggregated data on factor levels (capital and labour), efficiency of sectors, and innovation and diffusion patterns.

Various other formal evolutionary models of growth have also been proposed (Silverberg et al., 1988; Conlisk, 1989). Conlisk worked with a probability distribution of the productivity of companies. The growth rate can be determined analytically, and turns out to depend on the rate of diffusion of innovations and the size of innovations as indicated by the standard error of the probability distribution of productivity. Silverberg et al. (1988) formulated a model that originates from the Goodwin model, which is built around a formalization of the Phillips curve. This curve describes the relationship between price inflation and unemployment. Modelling a population of companies and their behaviour as fixed rules generates industrial dynamics. New capital results from profit, based on the rule that relatively profitable forms of capital accumulate relatively quickly. This can be regarded as a combination of capital accumulation and selection, in the sense that a technique with a relatively high fitness diffuses quickly through a growing population of technologies, as a result of capital accumulation. Companies can apply two different strategies to innovate: mutation or imitation. The probability of imitation by a company depends on the difference between its profit and the maximum profit in the population of companies. This follows the general model of innovation and imitation as developed by Iwai (1984).

Empirical research into diversity is often based on statistical analysis of differences between countries (Fagerberg, 1988). Important indicators used are input-related measures such as R&D expenditures and output-related measures such as patents. By combining these indicators with levels of productivity (income per capita), country clusters can be determined. Resulting insights include the following: R&D and patents turn out to have low correlation with productivity; R&D offers no guarantee of successful patents; and the growth rate can be inversely proportional to levels of productivity in the same period. This last insight implies a sort of catch-up mechanism, meaning that technical differences can be overcome by imitation. The general Schumpeterian non-equilibrium approach emphasizes the interaction between opposing forces: innovation that increases the technical differences between countries, and imitation or diffusion that leads to reduction of such differences.

2.3.2   Evolutionary versus Endogenous Growth Theory

This section describes briefly the principal similarities and differences between evolutionary and endogenous (neoclassical) economic growth theories. The reason for such a comparison is that both theories explicitly focus on the fact that growth is stimulated by technical change and that this in turn depends heavily on the structure and characteristics of the economy. The comparison is also intended to clarify the essential elements and the unique character of evolutionary growth theory.

Both theories endogenize technical change by treating R&D expenditures as a key variable. Neoclassical economic theory defines R&D at an aggregate or macro level. This is linked to the fact that neoclassical economic models are crucially dependent on an aggregated production function. Evolutionary theory directly derives both production and technical change from information about the company population, so that R&D is explicitly described as a phenomenon at the level of or within companies (micro level). Because of the population approach, evolutionary models are much better capable of handling diversity of behaviour and technologies, whereas neo-classical economic theories take representative or identical agents as their starting point. This latter approach occurs explicitly in a reasoning that is often applied within endogenous growth theory; namely, that production functions at the micro level can be replicated perfectly (since identical companies are involved), resulting in constant economies of scale at an aggregated level.

Furthermore, evolutionary models assume bounded rationality in companies, usually in the form of routines and learning by imitation. Neoclassical economic models assume perfect rationality, leading to the familiar optimal marginal decision rules. This explains the emphasis by neoclassical economic growth theory on growth-in-balance (or optimal growth), as opposed to the non-equilibrium features of evolutionary growth. The approach by Aghion and Howitt (1998) contains several elements of heterogeneity and destructive innovation – 'creative destruction' as coined by Schumpeter – in a neoclassical economic model type and thus sticks to the assumption of the rational agent. Mulder et al. (2001) call this a 'neoclassical Schumpeterian approach'.

Both theories can handle aspects of uncertainty and irreversibility, although this is more common in the evolutionary models. Evolutionary theories describe a specific type of irreversibility; namely, path dependence (see Section 2.4.6). Neoclassical economic growth models have difficulty describing the fundamentals of this phenomenon because of their aggregation level. A correct description calls for a population model, in which the distribution of characteristics of companies or technologies follows a historical path, such that this distribution changes irreversibly (Arthur, 1989). In addition, stochastic elements are not uncommon in evolutionary models, especially those aimed at identifying the inherent uncertainty about the type, scope and timing of innovations.

Neoclassical endogenous growth theory focuses attention on public externalities in technological innovation: the idea that knowledge is available to everyone without any rivalry being involved. Evolutionary theory emphasizes the barriers and delays in the process of diffusion of innovations, as well as imperfect replication and diffusion (Mulder et al., 2001). This latter aspect is recognized as an important source of innovation.

As illustrated by Nelson and Winter (1982) and Conlisk (1989), evolutionary growth models can generate patterns that link up closely with patterns that are generated by neoclassical economic models. Evolutionary growth models, however, can describe phenomena that are directly sensitive to government regulation, such as the rate of diffusion (imitation), company-specific innovation factors, selective forces and lock-in. They are thus capable of providing information about a broad range of policies aimed at welfare growth and technological progress.

It is good to note that evolution does not always imply growth, and vice versa. To understand the first point, we might consider an evolutionary process in which the diversity of companies changes while total output (in monetary or physical terms) remains constant or even drops. That growth does not imply evolution either is illustrated by a perfect replication – unlikely, though not impossible – of an existing productive activity, which will give rise to an increase of total output, while diversity will remain unaffected.

Since it ignores structure and structural change, neoclassical economic growth theory is not really capable of handling a time horizon that extends beyond a few decades. Stiglitz (1997) has argued that the time horizon of growth theory can best be considered to be around 60 years, thus roughly comparable to a major wave in a Kondratieff cycle. Since evolutionary growth theory can address structural change, it is in principle suitable to answer questions that relate to much longer time horizons.

2.4   KEY CONCEPTS IN EVOLUTIONARY ECONOMICS

2.4.1   Introduction

This section describes the key concepts that characterize evolutionary economics, as it arises from the line of thought elaborated in Sections 2.2 and 2.3. These concepts serve as a guideline for the analyses in the following chapters. Six key concepts are identified: diversity, innovation, selection environment, bounded rationality, path dependence and lock-in, and co-evolution (see Table 2.1). Each of these concepts is explained further in the subsections below.

2.4.2   Diversity

Diversity is the key concept within evolutionary economics and the prime element that distinguishes it from neoclassical economics with its basis in representative agents. The concept of diversity requires adopting a population approach, where heterogeneity prevails instead of representation.

Table 2.1   Key concepts in evolutionary economics

Diversity
- Companies (type, size)
- Techniques (production)
- Products (features)
- Strategies (marketing, R&D)

Innovation
- Combination/cross-fertilization
- Serendipity
- Education
- Isolation (spatial, economic)
- Co-operation
- Venture capital
- Niche markets
- Futures (future visions, scenarios)

Selection environment
- Physics (e.g. thermodynamic boundaries)
- Technology (technical feasibility, costs)
- Geographical features (including soil, water, wind and sun)
- Business features (organizational structure)
- Market (relative prices, market power)
- Institutions and public policy
- Specific conditions that affect R&D

Bounded rationality
- Time horizon
- Routines
- Imitation (diffusion)

Path dependence and lock-in
- Irreversibility
- Increasing returns to scale (economies of scale, imitation, learning and positive network externalities)
- Lock-in
- Extended level playing field

Co-evolution
- Subsystems
- Negative or positive feedback
- Spatial

In line with evolutionary biology and recent discussions about biodiversity, Stirling (2004) states that diversity is a multidimensional concept, characterized by variation, balance and disparity. Variation refers to the number of different technologies, processes, products, organizations, institutions or strategies in a population of elements. Balance or equality relates to the extent to which one or more elements dominate in a population in terms of size or number. Disparity refers to the degree of difference between elements in a population. These three dimensions of diversity each affect innovation and selection and thus future diversity. This can give rise to a development towards better performance (for example, in terms of welfare or reduction of environmental pressure) and increased complexity. This development has no specific focus, but it can head in a certain direction through steering by individuals or government action. Such steering, which is discussed in further detail in the next chapter, can focus on specific selection factors or on the stimulation of specific innovations.

Improvement of the performance of selected units is sometimes expressed through the concept of fitness, a measure for survival and reproduction. Fisher's theorem is relevant in this regard: 'The greater the genetic variability upon which selection for fitness may act, the greater the expected improvement in fitness' (Fisher, 1930). This insight can be associated with both innovation and selection. Greater diversity, after all, means more possibilities for creative combinations, while selection (survival and reproduction) that builds upon a broader basis (greater diversity) also leads to higher performance and thus fitness. The conscious pursuit of diversity thus makes sense. The goal must not be selection of the best in the short term, but stimulation of diverse portfolios. This implies that decisions based on a short-term cost–benefit analysis, thus aimed at efficiency or cost-effectiveness, may be detrimental to improvements realized by evolution in terms of higher welfare and lower environmental pressure in the long term.

An implication of the notion of diversity is that waste is inevitable in evolutionary processes. In other words, progressive evolution will usually involve the loss of elements that have a low fitness. This implies that innovation is not possible without wasting energy, effort and time, including the possibility of 'dead ends'. From an evolutionary perspective, however, these are to be regarded as necessary, inevitable features of the evolutionary process rather than as failures that should be avoided.

2.4.3 Innovation

Innovations can be categorized on the basis of a large variety of characteristics. The following is a non-exclusive list of characteristics:

- Products, processes or services.
- Factor saving or quality improving.
- Relevant to a single sector, various sectors or the whole economy.
- Codifiable/documentable or internalized ('tacit/embodied').

- Incremental versus radical.
- Public or private or a combination of the two.
- Unintended or organized.
- Autonomous or systemic.

Innovations come about in various ways. These can be related to the distinction between incremental (or gradual) changes, such as efficiency improvements in existing technologies, and radical changes, such as totally new concepts, technologies or products. Incremental changes can in turn take on two different forms. Isolated mutations can result from serendipity; in other words discoveries or inventions without any conscious search being involved, through a combination of chance, intelligence and knowledge or expertise (Fine and Deegan, 1996). Good education stimulates serendipity. In addition, incremental changes often result from processes involving inevitable, even unintended learning ('learning by doing') and specific search efforts (research by universities, public institutes and companies). Radical innovations often come about through cross-fertilization and combinations of existing concepts, technology or organization that have not been tried out earlier. The invention of the windmill in the 12th century can be viewed as a successful (re)combination of, on the one hand, gear technology developed for the water mill and, on the other hand, the sail that served to convert wind into dynamic kinetic energy (Mokyr, 1990, p. 44). The resulting windmill spread very quickly across Europe, stimulating a growth of energy availability for various types of economic activities, such as wood sawing, grain milling and paint production. Moreover, energy supply became thus independent of the geographical patterns of rivers and streams. Combinations (and recombinations) generally fare well with variation, implying a necessary level of 'wastage'. Wastage should therefore be encouraged, for without wastage there can be no innovation. Too little wastage can mean too little innovation.

It should be realized that radical innovations are disruptive to the socio-technical system and thus often face significant barriers, because the current system is not ready for them. They also face competition from sustaining innovations that build upon existing knowledge and interests.

Systematic search (R&D, science), much experimentation and collaboration can raise the chance of innovative combinations. In this context, the method of devising extreme scenarios and backcasting from them can make sense. The use of scenarios can fulfil a practical function in the development of insight into the potential variety in combinations of technologies, organizations and institutions. Many informative details about these matters lie with individuals but are not sufficiently integrated into a synthesis that might serve as a feasible alternative or scenario. Combining the knowledge that individuals, each with their own expertise, possess can stimulate and inspire, and can generate a new line of thought about previously unconceived combinations of elements. Thinking about extreme futures can inspire new combinations and other innovations.

A certain degree of isolation is useful, as it allows unique paths to be trodden that remain outside the influence of dominant technologies and thinking patterns. Such isolation may have a spatial dimension. Iceland, for example, might, because of its location, more easily take up a unique path in the direction of a hydrogen economy. Cultural differences between isolated regions contribute to unique approaches to innovation. Spatial isolation amongst regions within a country can also lead to such an effect. In addition, non-spatial isolation can be a source of unique developments. Niche markets are an example of this.

Major innovations are usually radical. The level of radicality appears to correspond to the level of uncertainty about the success of the innovation efforts. This type of innovation may be developed within fundamental research at universities and public research institutes. Alternatively, it may result from companies collaborating with each other or with government agencies. At a business level, venture capital is needed in order to enable radical innovations. This can relate to entrepreneurship in small companies as well as to research departments in large corporations.

The somewhat abstract and sometimes spatial interpretation of an innovation system has been proposed as a context to stress the fact that many factors are involved in the effectuation and success of innovations. These factors have been identified by various researchers. In this context, Porter (1990) referred to a 'growth diamond'. This diamond covers four dimensions: factor conditions, domestic demand factors, business strategy and business network. In addition, a spatial approach, with regional or national coverage representing a fairly homogeneous culture, is often implicit in the use of the concept of an innovation system.

An innovation system must not be confused with a linear system of innovation (Boschma et al., 2002, chapter 7). Innovations involve different phases. Distinctions are often made between fundamental and applied research and between invention and adoption. During the application phase, product and process development and diffusion are important phases. In real life, however, it is difficult to draw the line between invention and innovation. Few inventions are so perfect that they can be commercialized without adjustment. This in fact leads to the adoption of an interactive model of innovation, in which fundamental research efforts (inventions) are influenced by market experience (innovations or adoptions).

It is not easy to identify factors that either stimulate or discourage innovations. An interesting distinction is made by Diamond (1997). He distinguishes between direct ('proximate') and fundamental ('ultimate') factors, which economists refer to as endogenous and exogenous factors. Ultimate factors include the environmental and spatial features that are fixed in the long term, such as climate and the geographical structure of the continents. The geographical factor that Diamond emphasizes is the direction of the axis of the continents: north–south or vertical in the case of the Americas and Africa, and west–east or horizontal in the case of Eurasia. The horizontal axis of Eurasia has meant a relatively large experimental space for agriculture, since the same climate zone covers a region with an enormous land mass.

2.4.4   The Selection Environment

Selection refers to the process or set of processes that lead to changes in the population structure by reducing and giving direction to diversity. While this is not often noted, selection is not a concept with a single definition. Instead, it serves as an umbrella term that covers a wide range of meanings. This is due to the fact that selection is often regarded as an ex post and aggregated phenomenon that finds its origin in various micro and macro processes such as competition, imitation and external disturbances (for example, an economic recession or a natural disaster). Imitation is sometimes, however, segregated from selection, with selection then being viewed in the strict sense of removal ('selecting away') of an individual from the population. We use the term selection here in the broader sense.

Within biology, vehement discussions have been held about what constitutes the unit or level of selection. A similar problem exists within an economic context. Technology and company (organizations) can be 'selected'. As such, there is not a single starting point for selection but various levels at which selection takes place (van den Bergh and Gowdy, 2003). In what is called punctuated equilibrium theory, higher selection processes are referred to as sorting, as opposed to differential selection or survival at the individual and genetic levels (Eldredge and Gould, 1972; Somit and Peterson, 1989).

Selection has various dimensions. We cannot therefore speak of simple optimization through selection. Implicitly, a choice is made from among various selection factors, which together constitute the selection environment:

1. Physics (e.g. thermodynamic boundaries).
2. Technology (what is technically possible, costs).
3. Geographical features (including soil, water, wind and sun).
4. Internal company features (organization).
5. Markets (including relative prices, and the market power of a major player).

6. Specific conditions that affect R&D.
7. Institutions and public policy.

These factors can be explained as follows. A fundamental selection factor is the extent to which physical boundaries (1) are reached. Thermodynamic efficiency, for example, is limited. This applies as much to the conversion of fossil fuels by combustion engines into useful kinetic energy as to the effectiveness of the transformation of solar energy into electricity by PV systems. Up to the present time, these theoretical boundaries have seldom been reached in practice. This means that at the moment other selection factors play a more significant role. The limited knowledge plus the costs of advanced technology (2) together form a major selection factor. If a 'technological idea' is realized but the technical application is not yet possible because of insufficient knowledge or technological development, the application will not get off the ground. Selection therefore already takes place in an R&D phase or even earlier. Geographical differences (3) in soil, water (rivers, tides), wind and the amount of sunshine play a role, both in the technical execution of certain projects (for example, with regard to renewable energy sources) and in the further development of a proven technology based on its acceptance by the demand side of the market. Characteristics of organizations (4), especially business enterprises, also determine the selection pressure. These characteristics are sometimes referred to as internal selection factors (as opposed to the external factors). If an organization is not flexible enough to respond to changes in its environment or is subject to an internal conflict, a level of tension arises, as a result of which the survival of the organization may be at risk. Internal selection in combination with a flexible and adaptive organization, on the other hand, can lead to a focused search for new ideas, technologies and products in order to realize an adjustment to changing circumstances.

Selection by markets (5) can be divided into several types, depending on the goal and the perspective: product market, intermediary and input markets, financial markets and the labour market. Selection by financial markets takes place through the assessment by banks and investors of the expected profit potential of companies and the consequent decision on whether to invest or perform R&D. Producers of final products and services experience selection pressure from the consumer side of the market. Producers of intermediary products, process technology and related services feel, on the other hand, selection pressure from other, purchasing companies. Successful innovation leads via higher profits, better quality or lower prices to a growing market share. Market selection generally appears to be associated more with products than processes. That is because process costs are often only indirectly related to product costs and even less directly to product prices. This applies in particular to multi-product companies, where overhead costs are often allocated to products in an arbitrary or strategic fashion. Governments, as market players, can obviously play a role in the selection by markets. The level of competition clearly plays a major role in the removal of alternatives (products and companies). Perfect market functioning generally produces the highest selection pressure.

Research (science, R&D) (6) has its own specific selection mechanisms. A significant mechanism is peer review of academic publications, notably articles in journals. In addition, there are competitive schools that allow group selection or selection at the level of groups to play a role (van den Bergh and Stagl, 2003). At the level of private companies, R&D is subject to selection as a result of competition and effects on costs and benefits by first-mover advantages (secrecy, learning curve) and patents.

Institutions and public policy (7) are not only able to affect the other selection factors but can also operate as a selection factor themselves. Important elements of such policy include competition law, safety regulations (general and labour safety), environmental policy, labour market regulations, social insurance schemes and international treaties. Policy-related, economic and environmental factors can vary across space, leading to diversity in local and national selection pressures. Selection pressures on economic activities also include local factors (neighbours, space, accessibility and education), regional factors (infrastructure, regulation, markets, labour market and recreation), national factors (culture, regulation and public goods) and global factors (climate, international markets and international treaties). The term 'institutions' further covers the way in which information is obtained from and is negotiated by lobbyists, politicians and technical or scientific experts. The Dutch Polder Model – institutionalized via consultation bodies and advisory councils – is a good example of an institutional setting that exercises significant selection influence.

The selection environment is variable over time. In fact, given that the selection environment will often also be characterized by internal diversity, the joint evolution of a system and its selection environment can therefore be viewed as a special case of co-evolution (see Section 2.4.7). Both market and non-market selection environments and mechanisms can change, for example, through market competition, and entry and exit. Selection based on government regulation is also subject to change. This is partly the result of interaction between public opinion (including ethical standards), the media and science. An example of this is the debate about the regulation of genetic modification of organisms.

Selection is not deterministic. This is a reason why 'survival of the fittest' is an improper description of evolution. Just like innovation, selection is stochastic in character. Potts (2000, pp. 95–6) phrases this as follows:

'Selection is a regime, selecting not the most fit or profitable or, in some way optimal entity, but selecting a range of tolerably well fit. Selection as a filtering mechanism favours not the fastest, but the sufficiently fast, not the most profitable but the sufficiently profitable.'

2.4.5 Bounded Rationality

Bounded rationality means that individuals or companies do not act rationally or optimally, as evidenced by research on the imperfect implementation of energy-saving options (Jaffe and Stavins, 1994; Velthuijsen, 1995). Companies usually act in a routine fashion, even in their activities directed at investment and innovation. This creates a certain degree of heterogeneity among companies as well as sectors, since the routines are based on each company's own history and experiences. Bounded rationality thus implies a great diversity of strategies: all sorts of rules of thumb, rigid habits, routines, imperfect learning processes, conflict resolution and (imperfect) imitation of others. Diversity resulting from bounded rationality is of a different order than the application of this concept in neoclassical economics. In the neoclassical approach diversity is exogenous, with different economic actors applying different behavioural strategies. The evolutionary approach explicitly introduces endogenous heterogeneity in the behaviour of the economic actors, by letting behavioural diversity change in response to selection and innovation.

Bounded rationality is thus an almost natural aspect of evolutionary diversity. Indeed, bounded rationality contributes to diversity, while selection that moulds this diversity leads to automatisms and routines. The latter reinforce or maintain bounded rationality.

Characteristic of bounded rationality is that companies work with a limited time horizon (myopia) and therefore do not invest in matters that might be necessary for long-term survival. Nevertheless, there is diversity among companies in this respect, in the sense of some companies paying relatively greater attention to investments that only provide a return in the long term.

Behaviour on the consumer side of the economy is also marked by bounded rationality, which takes the form of habits and imitation of others (the latest fashions). The success of certain consumer products is very vulnerable to social interaction among consumers. In addition, information provided through the media and education can have a great impact. Within behavioural economics, which leans heavily on insights from economic experiments, much attention is nowadays paid to the fact that groups and social interaction have a large impact on the behaviour of individual economic agents. Here a number of social ('non-selfish' or 'other-regarding') preferences and emotions play a role: reciprocity, altruism, jealousy, hate, spite and aversion to inequality. This has consequences for policy theory. For example, the design of incentives changes when bounded rationality in combination with social preferences is considered (Fehr and Gächter, 1998; Fehr and Fischbacher, 2002).

2.4.6   Path Dependence and Lock-in

The concept of path dependence has received much attention in the neo-Schumpeterian literature about technological evolution. It reflects the idea that changes in heterogeneous populations are marked by increasing returns. These arise from various phenomena, such as economies of scale, 'learning-by-using', imitation on the demand side (fashions), network effects (such as in telecommunication), information externalities (when the market share of a product is large, its familiarity among potential buyers is large too) and technical complementarity (Arthur, 1989).

Increasing returns play a key role in the competition between alternative technologies. Whoever happens to succeed in gaining a greater market share is in the front seat and can grow relatively fast, to the detriment of competitors. The paths to a system configuration of a system then become relevant. This is typically the field of study of evolutionary economics.

A consequence of path dependence (or increasing returns) is that inefficient or unwanted equilibria may come about that cannot be escaped due to the increasing returns. This is referred to as lock-in. Coincidental historical patterns and events ('historical accidents') and starting conditions can have a major impact because of a high level of instability and sensitivity. Evolutionary systems with great diversity are relatively likely to follow a path-dependent process, since a high level of diversity assures that a state attained earlier is unlikely to be revisited. If diversity decreases due to the increasing returns of a dominant alternative, this can lead to a lock-in characterized by minimal, or even a total lack of, diversity. Lock-in means that it is very difficult to escape from such a situation, since the increasing returns then operate fully to the benefit of the dominant alternative. Moreover, the loss or lack of diversity that characterizes a lock-in situation implies limited opportunities for alternative evolutionary developments.

There is much empirical evidence that lock-in caused by increasing returns is more than a mere theoretical possibility. Well-known examples of locked-in – and suboptimal – technologies are the QWERTY keyboard, the VHS video system, the fossil fuel engine and the Windows® operating system. Even the fundamental dependence of energy supply and transport on fossil fuels can be viewed as an example of lock-in. Enormous economies of scale, network effects (think of petrol stations) and technical complementarity are at work here. In addition, R&D in car and electricity generation industries is strongly focused on improving technologies based on fossil fuels.

Lock-in of a desired or socially optimal system may seem something that one cannot object to. A drawback to such a 'positive lock-in', however, is that such a system may in time become less attractive as a result of a changing socio-economic context. Lock-in and an associated lack of capacity for evolutionary adaptation due to no or little (technological) diversity would then prevent a transition to an improved system. Given uncertainty about environmental (including economic) conditions, it can never be fully known beforehand when a lock-in is optimal. A certain minimum level of diversity of technologies and organizational structures keeps a system adaptively flexible, thereby guaranteeing an easier transition to a new, desired situation. This corresponds with a general evolutionary principle; namely, that diversity pays itself back as it contributes to fitness improvements. Systems that are able to maintain and generate new diversity allow for better performance in the long term. This line of thought suggests that, for example, in the context of energy supply one should strive for diversity in many respects, including energy carriers, renewable energy sources, and even a combination of centralization and decentralization of electricity production.

2.4.7 Co-evolution of Subsystems

Co-evolution is a concept that originates from the synthesis of ecology and evolutionary biology (see Box 2.1). It denotes that an evolving system exercises selection pressure upon another evolving system, thereby stimulating changes in diversity within it, and vice versa. The result is that the evolution of the one system interacts with that of the other. This distinguishes co-evolution from mere dynamic interaction among non-evolutionary subsystems of a larger system ('co-dynamics'), even though it is often misused as a catchword to refer to the latter (van den Bergh and Stagl, 2003; Winder et al., 2005). Nowadays, co-evolution is used to describe and explain a variety of interactions: biological/genetic–cultural, ecological–economic, production–consumption, technology-preferences-institutions (Lumsden and Wilson, 1981; Norgaard, 1984, 1994; Durham, 1991; Gowdy, 1994; Feldman and Laland, 1996; Wilson, 1998). Campbell (1996, p. 569) notes that the rise of agriculture with the domestication of animals and plants and the ensuing cultural and economic developments can be regarded as a special form of co-evolution between animals and plants. Human beings are dependent on cultivated and selected plants, and plants in turn are dependent on human control and management. In other words, the one cannot exist without the other. Norgaard (1984) was the first to apply the concept of co-evolution to socio-economic systems. He regards co-evolution as the set of long-term interactions between five subsystems; namely, knowledge, values, organizations, technology and environment. Variation in each subsystem is heavily influenced by selection pressure exercised by the other subsystems, which together constitute the overall environment of the subsystem in question. He illustrates his framework with the interaction between pests, the production and use of pesticides, policy and institutions to regulate pesticides, and knowledge and assessment of pesticides and pests.

BOX 2.1   CO-EVOLUTION IN NATURAL SYSTEMS

Co-evolution is a concept that was originally introduced in ecology to describe the joint and interactive evolution of butterflies and flowering plants (Ehrlich and Raven, 1964). Co-evolution means that evolutionary change in a particular species corresponds with changes in another species. Originally, the term co-evolution was used at the level of species; in particular, to explain reciprocal evolutionary adaptations, such as those of parasites and their host, of predators and their prey, and of herbivores and the plants that they eat.

Co-evolution must not be confused with the biological concept of co-adaptation. The latter refers to the simultaneous evolution of different characteristics – morphological, physiological and behavioural – within a species. This coincides with possibly complementary changes in information at various places in the genome due to selection. Co-evolution instead relates directly to ecological adaptation, since the co-evolving species are part of each other's environment.

In principle, two basic strategies are conceivable for the growth of populations and ecological systems. These strategies are traceable to the familiar logistic curve for population growth:

$$\frac{dx}{dt} = rx(1 - x/K)$$

Here x represents the population size and dx/dt the change therein over time. The strategies link up with the parameters for intrinsic growth (r) and for the carrying capacity (K):

- The *r-strategy*: produce as much offspring as possible and provide little parental care. This is typical for insects and amphibians, which are small, quickly become sexually mature and have a relatively short lifespan. Annual plants also fall into this category.

- The *K-strategy*: produce few offspring and provide much parental care. This is typical for birds and mammals, which are relatively large, reach their sexually mature age slowly and can become old.

In the light of these dynamics the growth curves of *r-* and *K*-strategists differ significantly. A population of *r*-strategists follows an exponential growth path, can overshoot its natural carrying capacity and then collapse, and can potentially generate cycles under unstable environmental conditions. *K*-strategists follow the logistical growth curve, which results in very stable populations (although the population can die out at low levels).

Young evolving systems are typically dominated by *r*-strategists, which are perfect settlers due to their characteristics. *K*-strategists are typically encountered in more developed or evolved systems, where a stable environment enables a more complex system of labour division and specialization that can support the *K*-strategists. The distinction between the two strategies is obviously a caricature, and the two can best be regarded as relative concepts. More or less extreme forms exist, both among and within groups of species. Nevertheless, evolution appears to have stimulated a movement from one extreme to the other (Putman and Wratten, 1984, chapter 10).

A specific form of co-evolution that has received much attention is what is referred to as the 'arms race', where characteristics of species react positively to each other (i.e. positive feedback). For example, fast predators select fast prey, and vice versa. The reason is that slow prey are eaten, while slow predators do not get enough food. Fast predators and prey will have more offspring, so that both species evolve so they can move at greater speed. Roughly comparable is the increase in the size of prey populations, so as to be better protected against predators. This in turn leads to selection of bigger predators. Lastly, the size of the brains of predators and prey is also subject to mutual selection, since this positively affects the hunting skill of predators and the skill of prey in avoiding predators (Strickberger, 1996, pp. 399 and 429).

Co-evolution at the level of socio-economic systems means that people do not control and adapt their environment in order to achieve preset goals, but that nature and human society are shaped in a joint and interactive development where goals may have a certain influence but are not ruling factors. Concepts such as progress and planning lose credibility in the context of co-evolution. They can better be replaced by concepts such as change and adaptation through experiment and selection. There are many examples of co-evolution involving socio-economic aspects (Boyd and Richerson, 1985; Durham, 1991; Ofek, 2001; Galor and Moav, 2002):

- The incidence of sickle-cell anaemia (a blood disease) among West Africans has turned out to be caused by the way in which these people support themselves. Cutting down trees is good for the cultivation of yams, but it causes pools of stagnant water in which mosquitoes thrive. These insects transmit malaria to human beings. Selection of sickle-cell anaemia then occurs, as this provides protection against malaria.
- The rise of agriculture with animal dairy products led to a selection environment in which the percentage of individuals with genes that allow the intake of lactose was able to grow.
- Life in densely populated communities with domesticated animals such as cows, pigs and sheep has resulted in the transfer of contagious diseases from these animals to humans and ultimately, after epidemics such as the Black Death (Bubonic Plague), to resistance against these diseases.

Durham (1991) provides a general classification of the co-evolution of genes and culture that may serve as a model for a more general application of co-evolution to the subsystems of an evolving system:

- *Genetic mediation*: genetic changes affect cultural evolution.
- *Cultural mediation*: cultural changes affect genetic evolution.
- *Enhancement*: cultural changes enhance natural evolution.
- *Opposition*: cultural change counters natural evolution (partially neutralizing its effects).

While this classification centres around genetic evolution, it can inspire a systematic elaboration of co-evolution to deal with broader issues of knowledge, collaboration, technology, economy and public policy. Even though accurate steering of a co-evolutionary system with general policy instruments seems impossible, one may anticipate feedbacks among subsystems and thus assess the potential dynamic patterns of the overall system. Co-evolutionary analysis aims to identify the wider system implications of this type of evolutionary feedback.

2.5 INTEGRATION OF THE EVOLUTIONARY CONCEPTS

This chapter has so far presented a brief introduction to evolutionary economics, a discipline with a long, but fragmented, history and a wide variety of concepts, specific theories and applications. The two major approaches are the neo-Schumpeterian school, which focuses heavily on technology, and evolutionary game theory, which builds a bridge between equilibrium thinking (neoclassical economics) and evolutionary thinking by focusing on repeated selection that results in equilibrium. In addition, there is an evolutionary growth theory that differs in various respects from the popular endogenous (neoclassical) economic growth theory, while they also share a number of characteristics.

Evolution is a concrete process, with diversity, innovation, selection and accumulation of features as the main elements. Innovation increases and selection reduces the existing diversity. Diversity, in turn, stimulates innovation, especially via combinations and recombinations. Selection is also dependent on the internal diversity of the population (consisting of individual units) in question. Finally, selected features accumulate to produce better adapted and possibly more complex units.

The distinction made in the previous section between the key concepts of diversity, innovation, selection environment, bounded rationality, path dependence and lock-in, and co-evolution represents an inevitable aspect of scientific reductionism, with the typical disadvantage that issues relating to two or more of these concepts are positioned in a somewhat arbitrary fashion. The reductionist approach also explains why certain matters are classified under more than one concept.

An attempt at a synthesis of the principal concepts leads to the scheme presented in Figure 2.1. The arrows in the scheme denote the impacts that an element or set of elements has on other elements. It should be clear that the resulting system, driven by evolution, is complex and anything but trivial. This means, among other things, that it is difficult to fully understand this system, and that accurate prediction or calculation of specific changes in it would call for a formal model approach. Intuition comes up short in this respect.

Evolution within an economic system relates to technology, organizations and institutions. It thus involves more than just technological variation. As to the behaviour of economic agents, bounded rationality is assumed. This expresses itself in the form of habits, business routines, imitation of others, and narrow spatial and time horizons. The non-rational behaviour of individuals and organizations implies a large variety of behavioural patterns since a representative or uniform rational agent no longer exists. Such diversity points once again to the relevance of evolutionary economic analysis.

Selection pressure must not be regarded as an unambiguous or one-dimensional concept, but as being based on a wide variety of selection factors. These include physical, technological, geographical, institutional, business-related and market dimensions. While the market dimension plays an overriding role in traditional economic theory, in evolutionary economics the entire range of selection factors receives balanced attention. These factors can serve as points of reference for policy that, by influencing particular selection factors, aims at steering developments in a desired direction. Policy can also exercise influence on selection directly; namely, through legislation and regulation in all sorts of fields (labour, environment, safety etc.). In this way, specific choices by economic agents are restricted or enforced. Thus policy implies changing and endogenous selection pressure. But selection pressure can also change autonomously, since the population changes over time due to innovation and repeated selection. At any rate, changes in selection pressure over time are inevitable.

Innovations occur in different forms. Incremental changes can be viewed as the result of isolated mutations, learning processes or focused search efforts (R&D). Radical innovations often come about through cross-fertilization or the previously untried combination of existing concepts, technologies or organizations. Serendipity can play a role in this. In addition, systematic search efforts (R&D, science), repeated experiments and collaboration can enlarge the chance of innovative combinations. A certain degree of isolation is useful, since it allows for new paths to be entered that are free from the influence of prevailing technology and thinking patterns. Niche markets are a practical elaboration of this. At the business level, venture capital is needed to be able to pursue such new paths. Combinations (and recombinations) fare well with variation. Wastage must be encouraged, for without it there can be no innovation, while too little wastage can mean too little innovation.

Evolution is path dependent and can result in a state of lock-in; that is, an evolutionary cul-de-sac. Path dependence means that early development phases exercise an irreversible influence on subsequent developments of a system. As a result, chance events ('historical accidents') in an early phase can heavily determine the characteristics of a system in the long run. In the jargon of traditional economics this means that there is more than one equilibrium, and that the ultimate equilibrium in which a system ends up is a matter of pure chance. To understand the attained equilibrium, and to design policies aimed at influencing or correcting it, insight into the paths leading there is essential. These are characterized by increasing returns to scale on the supply and demand side or, in less technical terms, by a

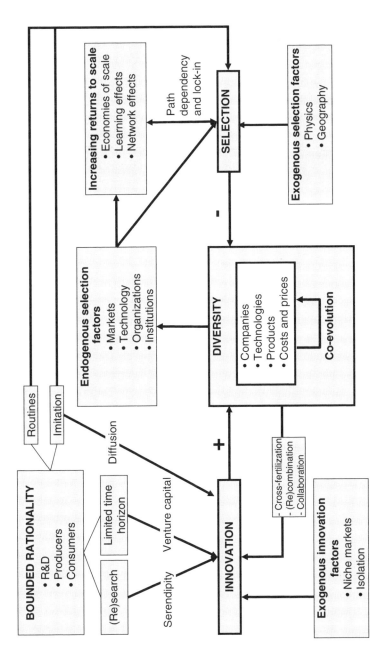

Figure 2.1 A synthesis of the main evolutionary concepts

self-reinforcing mechanism. A potential and likely result of this is an arbitrary and thus often inefficient or otherwise unwanted equilibrium, a lock-in from which escape is difficult if not impossible.

Lastly, co-evolution of systems can yield a useful concept for understanding complex systems dominated by interaction among subsystems and evolutionary mechanisms. Co-evolution is not simply to be interpreted as interaction between systems ('co-dynamics'), but as interaction or interdependence of evolutionary processes within two or more systems. Many inspiring ideas about mechanisms and consequences of co-evolution can be found in the literature. Co-evolution may be regarded as a dynamic elaboration of the notion of innovative (re)combination, where complementary, interactive subsystems change over time so as to become better adjusted to one another.

3. Environmental policy and transition management from an evolutionary perspective

3.1 INTRODUCTION

The evolutionary concepts and insights described in Chapter 2 will be translated in this chapter into suggestions for environmental policy and transition management. To illustrate the relevant principles and instruments, we will focus in particular on the transition to sustainable energy. The complexity of the energy system means that a great variety of scenarios can be sketched on the basis of any relevant combination of elements, such as degree of (de)centralization, energy source, energy carrier, technology and the extent of system integration. Through backcasting, the paths leading to these static scenarios can be derived, which in turn suggest policies that set adequate conditions for innovation and the selection environment.

The notion of 'transition' has its origins in population dynamics, so it ties in well with the evolutionary concepts that are used in this book (see also Kemp, 1997; Rotmans et al., 2001; Geels, 2002a, b). A social transition can be defined as 'a gradual and ongoing process of social change, leading to structural changes in society (or a complex subsystem thereof)' (Rotmans et al., 2000). Transitions thus actually represent system changes. They closely relate to system innovations as perceived in the literature on technological change. Various features and classifications of social and technological transitions frequently appear in the literature on sustainability and technological change.

From a temporal perspective transitions can be said to consist of the temporal phases pre-development, take-off, acceleration and stabilization. In addition, most transitions comprise changes at various spatial or system scales. Here a common distinction is between macro (landscape), meso (regime) and micro processes (niches). Next, one can distinguish between major and minor transitions. Examples of major transitions are the invention of how to light and sustain fire, the rise of agriculture and sedentary communities, and the Industrial Revolution. Minor transitions include changes within each of these regimes, such as the transition from coal to natural gas and the Green Revolution in agriculture. The transition to widespread electrification and the rise of the car are probably on the borderline between major and minor transitions. As to more recent changes, such as in the field of ICT or gene technology, their full impact and whether these involve major or minor transitions is not yet clear. Finally, the distinction between autonomous and intended or even managed transitions is relevant. The Industrial Revolution is often used as an illustration of the former, while the switch from coal to natural gas in the Netherlands is exemplary for the latter. The transition towards a sustainable development is an intended transition with a concrete goal (sustainability), which can be implemented in different ways. This type of transition is also often referred to as 'industrial transformation'.

Transition policy or management can be described as the stimulation and management of learning processes, keeping various options open, a multi-actor and multi-domain approach, and being motivated by a long-term interval goal (Rotmans et al., 2000). Transition policy is thus much broader in scope than traditional policy approaches in the fields of environment, energy and technology.[1] Transition policy encompasses elements of all these policy fields and involves technology policy, development of knowledge at individual and public levels, behavioural change, and alterations in organizations (including networks) as well as institutions (including markets). All these aspects integrate different spatial and temporal scale, from local to international. The role of the government is often regarded as one of guiding and facilitating these changes, rather than planning them top-down.

The policy approach described in this chapter differs from the neoclassical economic approach of public environmental policy (summarized in Box 3.1). In the first place, dynamic diversity, including the three dimensions of variety, balance and disparity (see Section 2.4.2), has no place in neoclassical economics. As such, it can say nothing about policy aimed at dynamic processes that result from interaction between innovation and selection, each of which influences and changes the three dimensions of diversity. An important insight of evolutionary economics is that diversity is the key driver behind evolutionary progress or benefits, rather than merely having a (short-run) economic cost – as emphasized by neoclassical economics. Evolutionary economics warns us not to select the best short-term option but, instead, to stimulate diverse portfolios that allow for evolution progress. Cost–benefit analyses that fail to account for the dynamics of diversity, as is common within traditional economics, will present an incomplete picture of the benefits. As a result, policy advice will deviate from (suggested) optimality.

Diversity is a concept that links up well with the idea of sustainable development. This applies as much to the concept of biodiversity as to cultural, economic and technological diversity. According to Stirling (2004) it is surprising that the study of sustainable development and analyses of diversity have not shown more interaction. An exception is the research on diversity and resilience in ecological economics (e.g. Gunderson and Holling, 2002). This suggests that diversification in a broad sense can be seen as a key concept in characterizing policies that aim to realize environmental sustainability.

Pursuing this line of thought, the other key evolutionary concepts mentioned in Table 2.1 also provide distinctive elements for an environmental policy theory based on evolutionary economics as compared with one based on neoclassical economics. These concepts include, next to diversity, certain aspects of innovation (notably recombination), selection environment, bounded rationality, path dependence and lock-in, and co-evolution. Section 3.3 discusses this more systematically. Bounded rationality means that assumptions regarding the behaviour of economic agents (companies and households) in traditional economic theory of environmental policy are not warranted and call for correction (for specific suggestions, see Section 3.6). As a result, notions such as optimizing behaviour and individual efficiency find little support within evolutionary economics.

To illustrate differences at a systems level, note that an important principle of the neoclassical economic theory of environmental policy is that prices must be correct, in other words that they should incorporate all social costs – the sum of private and external costs. An important generic insight of evolutionary economics is the idea that although correct prices are necessary, they are insufficient to escape from evolutionary states of lock-in (see Section 2.4.6). This suggests that, to a certain extent, the two approaches present complementary insights.

BOX 3.1   THE NEOCLASSICAL-ECONOMIC THEORY OF ENVIRONMENTAL POLICY AND TRANSITIONS

Theory of Environmental Policy

The economic theory of environmental policy starts out from the concept of 'negative externalities' or 'external costs'. These can be defined as the unintended impact of a decision by one economic agent on the welfare or production (and thus profit) of another agent, where this impact occurs outside the market (and is thus a direct physical effect) and remains uncompensated (Baumol and Oates, 1988). The presence of externalities – which is a rule rather than an exception due to thermodynamics and scarcity of space – implies that individuals do not have full control over all factors that influence their production or welfare. The field of environmental economics focuses on negative environment-related externalities, such as the negative physical effects of environmental pollution and the unsustainable use of natural resources.

It is good to realize that the concept of an externality is not unique to neoclassical economics and does not require a neo-classical economic analysis. However, externalities have mostly been examined analytically, using partial and general equilibrium theories from neoclassical economics. These theories assume rational behaviour on the part of individuals and companies. External effects disrupt the operation of markets and lead to results that deviate from optimal social welfare. Analysis of eco-nomic instruments of environmental policy is principally based on comparative statics and static equilibria. Change processes only receive sporadic attention.

Instruments of environmental policy are traditionally assessed on the basis of efficiency criteria, such as optimal social welfare, Pareto efficiency and cost-effectiveness. Effectiveness (the degree of certainty that the intended result is achieved) and distribution aspects (equity, fairness and justice) come second in line. The most common (archetypal) comparison is between uniform standards and levies to combat pollution. Levies have the advantage that they provide more efficient incentives than standards in the case of het-erogeneity (diversity) of abatement options and costs of polluters. They lead to either higher social welfare or lower costs of specific emission-reduction goals. The best-known pricing instrument is the Pigouvian tax. This is equal to the marginal extra costs of pollution in the (hypothetical) optimal equilibrium. Standards ('command-and-control') are especially attractive under extreme external costs (e.g. radioactive or toxic pollution). An instrument that combines properties of standards and levies is a system of marketable rights. This has two characteristics: a ceiling is imposed on all polluting emissions of a specific type by allocating a limited quantity of emis-sion rights; and these rights are marketable. The first characteristic assures effectiveness, as there is an upper limit to total emissions (at national, regional or sectoral level). The second characteristic guarantees flexibility at the level of individual agents, realizing equal marginal costs of pollution abatement among all polluters (such costs being equal to the equilibrium price of a pollution right). In addition to pricing instruments and standards, it is also possible to allocate ownership rights (the Coase solution). Assuming strict conditions (few polluters and victims, cheap access to the legal system), this can be an effective and efficient solution. Coase's main insight was that, given a number of conditions, the efficiency of the outcome turns out to be independent of the initial allocation of the rights to polluters and victims.

It is questionable whether neoclassical environmental policy theory has had much impact on environmental policy in practice. Although economists have pleaded since the 1970s for the use of market-based instruments, their recommendations have hardly been followed (Opschoor et al., 1994). There are two main reasons for this. First, political decision makers are very sensitive to specific lobby groups (Dietz and Vollebergh, 1999). Second, a country's international competitive position would be at risk if the policy advice were to be implemented unilaterally.

A Neoclassical-economic View of Transitions

What does traditional, neoclassical economics say about transition policy? It speaks of elements that are needed to realize a level playing field. These include inclusion of the (marginal) external costs of environmental pressure in relevant product prices, addressing the positive external effects of innovation (e.g. through patents or subsidies) and ensuring sufficient competition.

Two specific elements of neoclassical economics shed some light on transitions. First, (general and partial) equilibrium analysis provides a framework that makes it possible to analyse sectoral shifts resulting from exogenous factors, such as technological change or the availability of natural resources, as well as the gradual replacement of old and obsolete technology by new and profitable technology. The analysis is addressed, for example, with vintage capital models. Neoclassical economic analysis is limited to shifts that occur via markets, costs and relative price differences. In addition, economic (exogenous and endogenous) growth theory is used to answer questions about technology, R&D and investments in human capital (knowledge and skills). The main drawbacks of these two approaches are that they assume rational representative agents, that they largely ignore lock-in (an exception is Hofkes and Gerlagh, 2002) and that innovations are only described implicitly and, in the worst case, as being exogenous to the economy (see also Section 2.3). Only endogenous growth models, in which new technology is described as the result of R&D investments, offer an improvement in this respect. Because of their high abstraction level, however, these models offer relatively little information about the content, qualitatively different phases and underlying factors of transitions (see also den Butter and Hofkes, 2006). Diversity of technologies plays no role whatsoever in these types of models.

3.2　EVOLUTIONARY-ECONOMIC ANALYSIS OF ENVIRONMENTAL POLICY AND MANAGEMENT OF NATURAL RESOURCES

While scarce, a number of examples can be found in the literature in which theory and models from evolutionary economics are combined with issues and problems in environmental economics. We briefly review these here. For more complete surveys, see van den Bergh and Gowdy (2000) and van den Bergh (2004, 2005).

Evolutionary-economic theory neglects environmental and resource dimensions. Recently, we have witnessed the start of a synthesis of elements of environmental and evolutionary economics. For example, major periods and phases in economic history cannot be properly understood without calling upon economic-evolutionary as well as environmental and resource factors. The field within economics that focuses attention on these factors, environmental economics, is dominated, however, by neoclassical economic equilibrium thinking (see Box 3.1). This applies to three core theories; namely, about choice of environmental policy instruments, monetary valuation of environmental changes, and the exploitation and management of natural resources. The recent acceptance of the concept of 'sustainable development' has led to somewhat more emphasis on a long-term perspective and can easily be regarded as an invitation to apply evolutionary perspectives, in particular, to study structural economic and technical changes in response to environmental regulation (Gowdy, 1999; Mulder and van den Bergh, 2001). Norgaard (1984) proposes to apply the concept of co-evolution, which originated in biology, to study the joint and interactive evolution of nature, economy, technology, norms, policy and other institutional structures (see Section 2.4.7). Gowdy (1994) combines a similar view with macro-evolutionary elements, which leads to representing economic evolution and interactions with nature and the environment as a process that occurs at multiple levels (see also van den Bergh and Gowdy 2003).

Nicholas Georgescu-Roegen (1971) stressed the irreversibility of long-term economic developments from the perspective of the entropy law in thermodynamics. In addition, he introduced the concept of 'exosomatic development' of humans, meaning the systematic use of instruments and technology as a step in the evolution of the human race to outgrow its biological, endosomatic (physical) constraints or shortcomings. He regarded three technical, exosomatic innovations as being crucial ('Promethean') to major changes in the method of economic activity throughout history: control over fire, agriculture and the steam engine (Mesner and Gowdy, 1999). Although his work is sometimes characterized as 'bio-economic', it does not fully count as evolutionary economics, since it lacks the specific evolutionary concepts of diversity and selection.

Kenneth Boulding (1966, 1978, 1981), working from a similar background, namely, economics combined with care for the environment, tried to change economic methods by applying ecological and evolutionary biological analogies, such as ecological equilibrium, ecological stability, homeostasis and population dynamics. Boulding (1966), just like Hayek, stresses that evolutionary economics should focus on change in knowledge. In this context, he also proposes an analogy with biology; namely, the distinction between genotype and phenotype (worked out in detail later by Faber and Proops, 1990). Potts (2000, p. 58) criticizes this emphasis as being ineffective, since it elicits the question 'What, then, is knowledge?' Boulding (1981) was the first to write extensively on economic evolution from a systems perspective.

Until now, only a few concrete models have established the link between economy, environment and evolution. Munro (1997) has added evolutionary elements to the standard problem of the exploitation of a renewable resource. He argues that exploitation not only influences the size of the resource but also its quality; in other words, its composition in genetic terms. Examples include: monocultures, and the use of pesticides, herbicides and fungicides in agriculture; the period in the year when fishing takes place (for example, when there are many young and thus small fish), and the use of nets with a certain mesh size in fisheries; the management of groundwater in ecosystems; fire control in nature reserves; and the use of antibiotics in medicine. The genetic-selective effects of resource use and destruction of habitats are also linked to concern about the loss of biodiversity. Munro has formulated a dynamic optimization model that is based on the idea that the use of insecticides increases the fitness of resistant insects relative to susceptible competitors in the population. Optimal use of insecticides depends on the evolutionary selective dynamics of the system. In comparison with this, the traditional optimal plan, which ignores evolution, is myopic: it leads to an excessive use of pesticides, which eventually backfires on the user. Noailly (2003, chapter 5) expands the model to co-evolution, describing a population of farmers who employ different pesticide use strategies, but who imitate others if they perform better in economic terms.

Much attention has been devoted to the risk of over-exploitation of common property from common-pool resources, such as certain fish populations in open waters. Although common property is often confused with open access, where over-exploitation is very likely to occur, the risk of over-use is also quite real with common property resources. The risk depends in particular on the type of regime and can therefore differ from one situation to the next. A fundamental question is whether it makes sense to respond to conflicts about scarce resources and excess exploitation through strict government regulation at a higher level (and from 'distance'), or instead to rely on endogenous or spontaneous formation of local user regimes. An evolutionary perspective can be applied to analyse the latter option. The underlying idea is that such regimes can only come about if they receive sufficient support from the individuals who participate in or are subject to them. In other words, a specific social norm must evolve. Many contributions to the literature on the subject suggest that externally imposed rules and controls can be counterproductive, since they reduce and destabilize co-operation and interaction or even totally destroy it. Instead, it is preferable to support a norm through communication among the users of the resource. This can be modelled as a local interaction type of evolutionary game. External regulation is only desirable when an effective system of monitoring and sanctions can be implemented, which is not always the case.

Evolution of norms is a type of self-organization that is still not entirely understood in its fundamental characteristics. The size of the user group, for example, appears to be relevant for the self-organization and stability of an evolved system, but it is not clear which specific factors influence the critical population size at which stabile co-operation or complex social organization and institutions will arise. Changes in specific parameters, such as the price of the output of the exploited resource, or rules that are imposed by an external regulator, can lead to instability of an evolutionary regime. A regime can also collapse when the exploitation technology becomes more productive (technical progress) or when the price rises (Ostrom, 1990). This can mean a transition to a new user regime. These problems have been examined with a broad spectrum of methods, including evolutionary game theory (Sethi and Somanathan, 1996; Noailly et al., 2003), multi-agent models (Noailly, 2003: chapter 4), laboratory experiments and empirical field studies.

The above examples give the impression that environmental (and resource) economists have already made various attempts to bridge the gap with evolutionary economics. However, these examples have turned out to be mere exceptions. In general, it can safely be asserted that environmental economics has largely ignored evolutionary economics, and vice versa.

3.3   AN EXTENDED LEVEL PLAYING FIELD

Various principles that are key to evolutionary-economic theory can be applied in the planning or evaluation of environmental policy. These principles correspond with the key concepts introduced in Chapter 2 – diversity, innovation, selection, bounded rationality, path dependence and lock-in, and co-evolution (see Table 2.1). Diversity has already been discussed at length in Section 2.4.2, while bounded rationality is covered in detail in Sections 2.4.5 and 3.6. The other key concepts are each briefly discussed here from a policy perspective.

Innovation has many faces. Often it results from combining objects into something new, such as the combustion engine which, mounted on to an originally horse-drawn wagon, eventually led to the car. Innovation is partly the result of coincidence plus insight and education (knowledge). It is thus important to have proper education that is explicitly aimed at inspiring innovative talent. Innovation is increasingly stimulated by focusing on market opportunities of new products and technologies. In addition, setting up large research laboratories means close collaboration between individuals with higher education, which enlarges the chance that complementary ideas come together in the form of innovations. Three other factors are also important:

1.   *Isolation*, whether spatial or otherwise, contributes to an independent development that allows discovery of new paths. Tension can obviously arise between collaboration and isolation. Policy must stimulate both, for example, by fostering collaboration at local, provincial and international levels, and by supporting protected niche markets.
2.   *Diffusion* or adoption of existing ideas or technologies sometimes leads to unintended errors or adjustments that in themselves constitute innovations. Policy measures must stimulate diffusion so as to create more space – both literally and figuratively – for experimenting.
3.   *Risk-seeking behaviour*, supported by venture capital, enhances the chance of innovations. Policy must create the preconditions for more risk-seeking behaviour, for example, through tax incentives and policies that are specifically aimed at investors and financial markets.

The *selection environment* determines, in addition to innovation and path dependence (see below), the development of an evolutionary system.

Governments can try to influence or even mould this selection environment, taking into account its dimensions as described in Section 2.4.4: physical, technological, geographical, business-internal, market and R&D related. Physical and geographical conditions can hardly be influenced by policy. The technological preconditions can be favourably impacted by ensuring that sufficient basic knowledge (education) and basic techniques (a sufficiently diverse sector structure) are present at the national level. As for internal business conditions, governments can influence these only in a minor way, although possibly indirectly through its impact on education and on conditions for entrepreneurship. Market circumstances can be partly controlled by governments, for example, through anti-trust legislation. R&D conditions depend on tax legislation, subsidy regulations, co-operation between public and private sectors (universities and institutes of technology) and specific innovation policies. All these factors combine to create an economic selection environment that is fairly complex and itself subject to constant change. With regard to this last aspect, recent trends such as liberalization, privatization and internationalization affect the selection environment in a major way. In the field of energy supply, for example, the stable selection environment of utility companies with a monopoly position now belongs to the past. Substantial selection pressure is generated by environmental regulation, especially that aimed at controlling the emission of greenhouse gases and acidifying substances, or aimed at investors ('green funds' for wind and solar energy) and consumers (such as 'green electricity').

With regard to *irreversibility and path dependence*, which options are successful or not depends to a large extent on coincidence rather than on performance in terms of efficiency, profitability and sustainability. The reason is that due to increasing returns to scale, a technology that happens to see its market share grow relatively fast in an early phase of development and competition with related technologies will not experience serious competition later on. This may be caused by many different factors, such as cost advantages from large-scale production, network effects, fashion or imitation effects, information externalities, standardization and complementary technology.

Lock-in of technology and organizational structures implies that once a configuration becomes dominant as a result of increasing returns to scale, it is very difficult to break out of it. Price corrections to incorporate the external costs of environmental damage in prices are a necessary but usually insufficient condition for undoing the lock-in of a system. The argument here is that competition among alternatives occurs on the basis of cost differences that can be significantly affected by increasing returns to scale. Additional policy measures are thus needed, notably stimulating alternatives through subsidies, creation of niche markets, setting clear long-run objectives, providing information, educating consumers and so on. In particular, unlocking will require, next to (environmental) regulation, some type of innovation policy. Possibly, an alternative approach is to 'overcorrect' prices, in the sense of charging beyond marginal external costs, so as to reflect not only environmental externalities but also the external costs associated with lock-in (Hofkes and Gerlagh, 2002). However, this is unlikely to receive much political and public support, and certainly not more than a combination of regulatory and innovation policies. We will see that from an evolutionary perspective a so-called 'extended level playing field' is required, in order to avoid early lock-in, to stimulate unlocking (i.e. once lock-in has occurred), and to maintain or stimulate technological, organizational or institutional diversity. The notion of the extended level playing field is characterized below.

Co-evolution is a concept that offers insight into the complexity of evolutionary systems. The reason is that not only does each subsystem possess its own variation, innovations and selection pressure, but also interaction takes place between these subsystems. This means that changes in the diversity of subsystems are interrelated. For innovations it is useful to investigate which technological and sectoral complementarities play an underlying role and to focus policy on stimulating them. In addition, co-evolution suggests that innovations must be able to lean as much as possible on a diversity of complementary technologies, instead of totally depending on a single complementary technology. The latter situation could, after all, result in an unwanted barrier or early lock-in.

The combination of all the previous elements shows a strong argument for the conviction that in order to govern a transition, covering the unlocking of existing structures, one or multiple connected system innovations and possibly the prevention of early lock-in, it is necessary to create a set of conditions that we refer to as an *extended level playing field*. Whereas a level playing field is generally regarded to be a condition under which parties can compete with each other on equal and fair terms, an extended level playing field must encompass more than just a 'free market' (i.e. the absence of an imperfect market). The following five additional elements are important:

1. *Prices* should reflect all (marginal) social costs; that is, the sum of (marginal) private and external costs.
2. Within the various technologies that are estimated to have the potential to contribute to sustainability goals, those that are at the beginning of the *learning curve* may be given a relatively high level of support, so that they can compete well with technologies that are already further

along the learning curve, or that are locked in but are less promising from a sustainability perspective.

3. The factors that determine the *time horizon* of investors or companies should be identified and then measures taken to equalize this time horizon as much as possible among all agents relating to competing alternatives (technologies).
4. Correction of the *increasing returns to scale* that favour one particular technology over another should be considered, in order to avoid undesirable early lock-in.
5. The differences in *selection environment* other than markets should be corrected, so that every alternative (technology or organizational system) is subject to the same selection factors.

In a metaphorical sense, we can say that the creation of an extended level playing field involves not only the removal of the molehills (= ensure good functioning of the market), but also levelling the slope of the field (= the five additional elements above).

Lock-in of a single technology should be avoided, even if this technology contributes to the sustainability goals. Lock-in always means the loss of flexibility and resilience of the technological system and it will counter the entrance of alternative and possibly even 'better' technologies in the future. Policy measures should therefore strive to accommodate a mechanism that ensures a certain degree of diversity.

What is the optimal level of diversity? This is not an easy question, but it should be addressed somehow. The cost of maintaining an extended level playing field with various options can be offset by the cost of loss of alternative options through early lock-in of a given technology. Conceptually, the trade-off between efficiency and diversity is the distinguishing, fundamental approach of evolutionary economics. It is pertinent to search for useful guidelines to be able to make this trade-off. In practice it usually involves trial and error, or in the worst case a focus on short-term efficiency, especially since an evolutionary conceptual framework is neglected. In most cases, the trade-off between efficiency and diversity cannot take place on the basis of a fully quantified analysis of costs and benefits, because of the inevitable uncertainties of moving along a learning curve. The comparison of the short-term costs with the long-term benefits of diversification represents a qualitative assessment that depends to a large extent on the opinions and intuitions of experts. A good method for carrying out this assessment is not available and still requires much research (van den Heuvel and van den Bergh, 2005).

Evolution is often viewed as a process without purpose, but there is more to it. Transition policy accepts the principle that system changes are largely driven by purposeless evolution, but it adds normative elements to the evolutionary system. These elements can be put into effect through policy aimed at stimulating innovation or influencing the selection environment. The result is autonomous evolution within boundaries, with the boundaries being adjusted from time to time in a politically desired direction. This idea of managed evolution is in essence not different from the traditional, neoclassical economic approach. The latter links autonomous, purposeless market functioning to policies aimed at correcting various types of market failure, so that markets are at the service of social welfare. Similarly, evolutionary change of the economy guided by adequate environmental and transition policies can serve the goal of sustainability. In comparison with a neoclassical economic approach of rational agents and stable market equilibrium, an evolutionary-economic view characterized by bounded rationality, lock-in and complex co-evolution may well provide a more sceptical picture on the extent to which socio-economic change can be affected by public regulation and management. This must be regarded, however, as a 'shortcoming' of reality rather than of evolutionary economics.

3.4 CASE STUDY: ENERGY SUPPLY AS A COMPLEX AND EVOLVING SYSTEM

The remainder of this chapter will carry on the evolutionary analysis of environmental policy and transition management with a special focus on the example of sustainable energy provision. The reason for this choice is that the generic insights obtained thus far can then be supported with concrete examples. By employing a well-bounded and firm context, one that moreover is at the core of a transition to a sustainable economy, we can generate a coherent and more complete view of environmental and transition policies.

An important message of evolutionary economics is the need to keep options open for long enough to allow promising (technological) options a chance to emerge fully. Which options and how long precisely they should receive support is not clear in general. We therefore cannot readily say in what exact way an ideal, low-emission energy system should be realized in the future. Not only does no one know the ideal system, but a rigid opinion combined with a deterministic view might well be counterproductive and lead to wrong decisions by both governments and companies. The complexity of the problem calls for a vision of the future that is not too narrow. Having said this, what is important is the exploration of possible future systems and the paths leading to them, since valuable policy lessons can emerge. The following parameters are crucial with regard to the context of energy provision:

1. *Centralized versus distributed (decentralized) energy supply*: generation of energy can occur with small-scale technologies at local, factory, residential or even individual home scales, as opposed to national or even international (European) scales with linked energy distribution networks and large power plants. The choice between these is closely tied to the presence of economies of scale versus flexibility and diversity.
2. *Energy carrier(s)*: in particular, the choice between electricity and hydrogen or some combination of these may be important in the future.
3. *Energy source(s)*: a broad range of alternatives is available, including fossil fuels, nuclear energy and various types of renewable energy, such as solar, wind, water, sea wave, biomass and geothermal energy.
4. *Technology*: various types of basic technology are combustion, nuclear fusion, nuclear fission, the fuel cell and so on.
5. *Extent of system integration*: this covers mainly traffic separated from or integrated with stationary sources and users; and extensive or limited use of combined heat and power (CHP) and residual heat.

These parameters are not independent. Notably, parameters 3 and 4 are closely linked.

Given the expected complexity of, and uncertainty regarding, the future energy system, it is important to keep multiple options open. Early choice of a specific strategy can lead to bad decisions from the viewpoint of intertemporal efficiency and long-run sustainability. In a policy context, strategies are often linked to criteria that the system must meet. The three criteria used to evaluate current policy are reliability or stability of supply, cost-effectiveness or price (notably of electricity, gas and gasoline), and environmental sustainability. The complexity of the energy supply system and the three partly conflicting criteria make it far from easy to test which future strategy is best. The pervasive uncertainty about an attractive energy system for the future is illustrated by the continuing research into nuclear fusion. This technology is far from ripe for commercial application and unable to compete with other technologies. But in the long run, more than 50 years from now, it may well be attractive in view of all three criteria. However, this will require a very substantial public financial investment by many countries, in addition to collaboration between public and private parties in terms of knowledge and technology development (see also Section 5.3).

Closely related to the degree of centralization is the issue of the extent of system integration. This involves, for example, the use of combined heat

and power, and the question of whether traffic/transport and en
to households and industrial consumers should be integrated. I
to assess all the possible consequences of such integration. But ı
the key to a sustainable energy system of the future.

Similarly, decentralization will have an influence on ma..., system
features, including generation of electricity, transport and distribution,
storage and consumption. An example of distributed generation is the cou-
pling of a car fuel cell to the house (e.g. during the night), implying that end
users of energy are at the same time producers. Parallel systems for industrial and private consumers are also a possibility, since the economies of
scale for these groups do not need to be identical.

The full range of economic and technological consequences of a fundamental change in the decentralization, integration or planning of energy
provision is difficult to judge. The costs of networks and the economies of
scale play a role, while another aspect to take into account is behavioural
changes in response to new technology, especially in the case of far-reaching
decentralization. The potentially unexpected character of such behavioural
changes is illustrated by the rebound effect of energy efficiency improvements, which covers a broad range of individual, market-related, sectoral
shift and other macro effects. For example, consumers may keep energy-
saving lamps turned on longer and more often than the traditional light
bulbs. It will be difficult to predict all possible behavioural (rebound) effects
of future developments of the energy system.

3.5   THE ROLE OF THE GOVERNMENT

The government is generally the party that needs to take the initiative in
setting up a form of transition management. This is due to a combination
of three factors: external environmental effects (social costs), external
benefits of innovation (knowledge spillovers) and lock-in of a socially
undesirable technology. In combination, these factors introduce a social
problem that cannot be resolved by companies or the market alone. Only a
leading role on the part of the government, complemented where necessary
by generic and specific policies, can solve this social problem.

An important consideration in transition management, which links up
with the concepts of evolutionary economics, is that the government, while
playing a major role in coordinating and influencing the transition process,
can never exercise full control over this process (Rotmans et al., 2000).
Nevertheless, the government can play a role in assessing scenarios, dealing
with uncertainties and lock-in, coordinating short-term policy with long-
term objectives, and dealing with the different scale levels (in time, space

and institutions) in which the transition takes place. A traditional, hierarchical governing style will increasingly need to be replaced by a more network-oriented approach, in which the role of government changes from ruler to manager (Faber et al., 2003). Important elements in shaping the role of the government in evolutionary transition management are the following (see also Faber and van Welie, 2004):

- Incorporating the (marginal) costs of external effects in prices through rules or levies and a stable pricing policy. Subsidies and green certificates can play a role here, but this is not enough.
- Improvement of information exchange between players, so that the collective learning process is more effective. Subsidies can play a role in generating new knowledge and technology.
- Assistance in eliminating or avoiding lock-in (keeping options open). Niches can be created, for example, through subsidies to consumers or producers. Niche markets can accelerate the learning curves and benefits of economy of scale. From an evolutionary perspective, stimulation of niches is a specific case of a general condition that stimulates major innovations, namely isolation. Lock-in is unavoidable to a certain extent, but policy can help ensure that such lock-in does not take place in technologies that are inferior in societal terms, and that room is continuously created for potentially superior technologies to break through such a lock-in. Government tendering policy can play a role in stimulating niche markets (the public sector as 'launching customer'). Stimulation of diversity runs counter to the operation of the market and causes extra costs. Here lies a task for the government.
- Protection of niche markets until the related technological developments are ripe enough to face the competition of the free market (see Kemp et al., 1998).
- The government plays an important role in supporting *scientific research* into fundamental innovations, since the payback periods and uncertainties make such research insufficiently attractive for private partners.
- To a certain extent this also includes scientific research into transitions, with a focus on theory, historical analysis and multidisciplinarity (technology and the social sciences). Such research can lead to lessons from past transitions that were unsuccessful, successful (e.g. the transition from coal to natural gas) or partially successful (e.g. the transition to a nuclear age) (see Geels and Smit, 2000).
- Collaboration between public and private parties in fundamental scientific research can be useful, although problems of secrecy and

rights to research findings may arise. Investments involving considerable risk fare best in the public domain.

- Government can play a role in influencing large players, especially when the market is dominated by a certain technology. Policy in this context may be aimed at achieving diversity, not only of technologies but also of the players involved. This comes down to a sort of portfolio management, meaning that a wide range of promising initiatives is stimulated, with alternatives that turn out to be less successful or promising being removed in due time (Kemp and Rotmans, 2004).
- Lastly, transition policy calls for constant adaptation to changing circumstances (external, exogenous factors) and processes that go in undesired directions. This may include the gradual elimination of initial support as part of strategic niche management. Clear choices must be made from time to time in both the positive and negative sense – in other words, providing stimuli or reducing support.

3.6 BOUNDED RATIONALITY

3.6.1 Bounded Rationality and Environmental Policy

If we let go of the assumption in neoclassical economics that agents focus on maximization of their profit or utility and always succeed in that goal, then the first and second welfare theorems of microeconomics no longer apply. This means that there is no direct relationship between a market equilibrium and optimum social welfare. For environmental policy theory this implies that there is no simple relationship between optimal externalities and environmental policy instruments, or no such thing as an optimal policy. After all, the idea that agents react optimally to marginal values and incentives no longer holds. Note, however, that the key notion of externalities continues to apply, since it does not require that agents optimize their profit or utility. A related conclusion must be that economic valuation – which assumes rational agents – loses relevance as it operates less accurately (van den Bergh et al., 2000).

A consequence of bounded rationality for instrument choice in environmental policy is that pricing instruments lose their appeal to some extent. Prices obviously impact choices, but pricing instruments become less effective and efficient than neoclassical economics would lead us to believe. More effective instruments then come in the picture as attractive alternatives, such as technical standards and physical regulation of emissions.

In the case of bounded rationality strategies such as 'satisficing' and 'routines', responses to environmental policy will be less evident than in the case of maximizing behaviour. Satisficing implies that, at best, cost-effectiveness is the aim. Tradeable permits then become relatively appealing. Habits and business routines provide a satisfactory explanation for the 'energy gap'; that is, the unreaped economic benefits of energy-saving options available to households and firms. Van Raaij (1988) clarifies this point in the context of consumer behaviour models that take into account such matters as visibility, demonstration value, environmental care, information, habits and socio-demographic determinants.

Roe (1996) applies insights from Girardian economics, a theory that describes economic responses to extreme uncertainty. It regards imitation behaviour as a logical response, which in turn leads to a reduction of economic diversity in terms of behaviour strategies, ideas, products and so on. This theory is applied to the general notion of sustainable development, viewed as a social convention that stabilizes decision making under extreme uncertainty for a period of time. It ought to lead to a trend of 'undifferentiation' (homogeneity) resulting from imitation being countered or reversed. This result can be realized, for example, by reducing uncertainty at the individual firm or consumer level, by buffering environmental systems, or by uncoupling them from general environmental uncertainties (such as climate). Another strategy is to promote diversity by stimulating multiple types of sustainable development, such as allowing for decentralized and thus unique initiatives; for example, at city or provincial scales.

The idea that preferences of consumers are endogenous and evolutionary (e.g. recombinant) instead of invariant led Norton et al. (1998) to the idea that changing consumer preferences can be a component of environmental policy. In particular, these authors asserted that stable preferences are at best realistic for short periods of time, while sovereign preferences can be inconsistent with long-term objectives of sustainability. In their view, a public discussion about responsible consumption and sustainability must be stimulated through, among others, education, rules for advertising and cultural norms. This is sometimes referred to as 'moral suasion'. Economists have for a long time accepted the view that decisions may be constrained by (public) regulation. In addition, preferences can thus be regulated, even if indirectly, such as by constraining commercial advertising. Most democratically elected governments have already implemented a number of public policies intended to combat actions that are considered criminal, racist or otherwise harmful (driving at excessive speed, smoking) to society. In economics jargon, 'harmful' here means that negative externalities are caused upon others. In principle, the previous list of actions can be expanded to preferences that conflict with sustainability (negative, environmental externalities). After all, we accept the fact that consumer preferences are heavily influenced by advertising, so why not that they are limited or directed for the common good? For a detailed discussion of bounded rationality, alternative behavioural models and environmental policy implications, see van den Bergh et al. (2000), and for a rare empirical–econometric application, Ferrer-i-Carbonell and van den Bergh (2004).

3.6.2   The Short Term as a Barrier

An important element of bounded rationality is the short-term horizon (see Section 2.4.5). This often operates as a barrier to the realization of long-term goals. Both individuals and firms are generally focused on achieving short-term results. A consequence of this is the tendency to discount, which results in more attention being given to short-term rather than long-term effects. This appears to be the case with energy conservation. While in the short run it does not seem to be inconsistent with transition goals, energy conservation can nonetheless be counterproductive, since it reinforces the lock-in of the present system. This is illustrated by the fact that many industrial countries are able to meet their CO_2 emission goals in the context of the Kyoto Protocol without fundamentally adjusting the method of energy provision. In other words, there has been relatively little pressure on the realization of sustainable energy use through renewable energy sources. The low CO_2 emission goals can be understood on the grounds of a short time horizon adopted by the parties in the international climate negotiations.

Myopia, that is, a short time horizon in decision making, can influence the choice between investments in energy conservation, renewable energy and carbon sequestration ('clean fossil fuels'), as the costs and benefits of these alternative options differ in the short run. Moreover, the pursuit of energy efficiency and conservation tends to reinforce the dominance and lock-in of fossil fuels and related technologies, since both their operational cost and the environmental pressure will be reduced in the short term (see Box 3.2). To prevent such reinforcement of lock-in, the striving for energy efficiency might be complemented by two strategies: (1) stimulating types and directions of improvements in energy efficiency such that reinforcement of fossil fuel lock-in will be minimal; (2) extra (compensating) investments in renewable energy sources (wind, sun, water) or alternative energy carriers (notably hydrogen) to counter relative cost gains for fossil fuel technologies due to energy conservation. A similar reasoning applies to combined heat and power (CHP), as well as to CO_2 capture and storage (sequestration), because these, like energy conservation, can reinforce the lock-in of fossil fuels. In the short run, it would further make sense to allow or even foster combinations of old and new technologies so as to stimulate the development and large-scale application of new technology and of related learning effects. An example is letting fuel cells run on hydrogen derived from fossil fuels (see also Section 5.2).

Concrete measures that can extend the short-term horizon of economic agents to a longer term might be directed at adjustment of interest rates, stimulation of venture capital and creation of niche markets. All of these options have a price tag and are in that sense partly interchangeable. Policy measures may also shorten the time horizon of economic agents instead of extending it, which is an example of a policy failure. For example, taxation that assumes a fixed return on investment provides a stimulus to invest in projects with a high yield in the short term. This will be at the expense of projects that yield a profit further down the road.

Citizens, politicians and other economic agents not only often employ a short-term horizon, but also often focus on the small scale at the expense of the large-scale perspective. In general, individuals – and consequently also organizations and even countries – tend to discount problems that are distant in terms of time and space. The end result is a deadlock or 'Prisoner's Dilemma' that is characterized by postponement of (necessary) actions and 'free riding'.

On the international scene, the limited horizon in time and space becomes evident in the liberalization of the energy market. The growing role of private parties strengthens the impact of short-term activities at the expense of long-term and large-scale investments (EZ, 2002, p. 58). This can have all sorts of negative consequences for stability as well as for the environmental sustainability of energy provision. An example of this is evident in California, where lagging investments in expensive production facilities have repeatedly led to interruptions in energy supply.

BOX 3.2   THE INFLUENCE OF ENERGY
            CONSERVATION ON A TRANSITION TO
            RENEWABLE ENERGY

Many types of energy conservation go hand in hand with a more efficient use of fossil fuels. The following general classification of energy conservation is useful: (a) a lower demand for energy through, for example, insulation of homes, on cars with less air resistance; and (b) more efficient conversion of primary energy resources, notably fossil fuels, into other forms of energy, such as heat, motion and electricity (e.g. condensing boilers, more economical engines). Energy savings arising from a lower demand do not necessarily lead to a change in the cost per unit of energy produced, except perhaps when a change in demand is so large that economies of scale are enjoyed. In the case of energy savings resulting from more efficient energy 'production' or conversion, however, the unit costs of energy produced will always go down. This mechanism leads to reinforcement of the economic position and thus the existing lock-in of fossil fuels. As a result, the ultimate transition goal, that of achieving a sustainable energy supply with major support for renewable energy, will come about more slowly.

To determine the effect of a delayed transition to renewable energy on total CO_2 emissions over time, two separate components must be determined. These are the emission reduction resulting from energy conservation over time and the avoided emission reduction due to a delay in the transition to renewable energy sources. These two components move in opposite directions in terms of size when energy conservation efforts are reduced or increased. This points at an optimal trade-off resulting in minimal total CO_2 emissions over time, even though such an optimum cannot be quantified accurately in view of the many uncertainties involved. The idea of a trade-off between investing in energy conservation and investing in renewable energy consideration is depicted in the following figure:

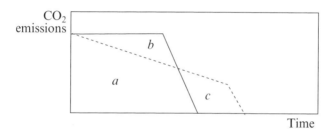

The continuous line in the figure represents CO_2 emissions resulting from a transition without or with just little energy conservation. The broken line shows the development of a transition that involves significant investments in energy conservation. In the latter case, the transition to renewable energy sources (not involving emissions of CO_2) starts at a later time.

The total emissions under the scenario without energy conservation are equal to $a + b$, while those under the scenario with energy conservation are represented by $a + c$. The scenario to be preferred thus depends on whether $b > c$ or $b < c$. The size of area b can be interpreted as the extra reduction of emissions through energy conservation and the size of area c as the missed emission reduction due to the delay in a transition to renewable energy sources.

If the time delay of the transition caused by energy conservation is very large, then the 'loss' *c* will be relatively large. From the viewpoint of cumulative CO_2 emissions, the scenario without energy conservation is then to be preferred. A larger investment in energy conservation implies that *a* will become smaller and *b* larger, but also that the transition will be further delayed, causing *c* to become larger. This means that more energy conservation and efficiency improvements are not necessarily a blessing for cumulative CO_2 emissions in the long term. This aspect deserves attention when analysing an energy transition. To the extent that energy conservation is more successful, the transition to costly renewable energy sources will be slower. Care should therefore be taken in trying to achieve a maximum degree of energy savings. This is even aside from the consideration that part of the investments in energy conservation may be applied to R&D in sustainable energy production, and that renewable energy will become cheaper if it is applied to a greater extent (learning curve effect). Last but not least, energy conservation will mitigate the required sense of urgency that a transition to sustainable energy sources is needed.

In fully understanding the above point, it should be realized that a truly sustainable energy system can never be realized through energy conservation but will, instead, require a large dependence on sustainable energy sources. After all, energy conservation alone will never lead to an energy system that is free from fossil fuels and CO_2 emissions. The transition to renewable sources is thus key to a sustainable long-run energy system.

The most important policy advice to be drawn from the above is that the advantages (direct reduction of emissions) and disadvantages (indirect increase of emissions due to delayed transition) of energy conservation must be carefully considered. In practice, such a trade-off is not easy, since the future is full of uncertainties. Empirical historical research is undoubtedly useful to be able to quantify to some extent the scope of the effects of energy conservation on the time path of a transition. Nonetheless, the following general conclusions about policy can be drawn:

1. Regulators should set limiting conditions on energy conservation, aimed at stimulating technologies that are not specifically linked to fossil fuels, but that are totally neutral with regard to fossil and renewable sources. The above distinction

> between energy conservation through demand and conversion may turn out to be useful in this regard.
> 2. If extra efforts (investments) are made to conserve energy, these should be offset by extra stimulation of, or investments in, sustainable energy production with renewable sources.
>
> A combination of 1 and 2 is possible as well, since these two policy options are not mutually exclusive.

3.7   DIVERSITY AND SCALE COMBINED: FROM LOCAL TO INTERNATIONAL

Transitions – in general or related to sustainable energy – are not limited to individual countries. It is almost impossible for a single country, especially if it is not very large, to initiate a specific transition on its own. International co-operation and coordination of transition policies therefore makes sense.

We have seen that evolutionary transition policy must find a proper balance between diversity of options and economies of scale (or more generally, increasing returns to scale). Until now the context was implicitly the national level, but in view of the advantages of a large scale it would appear easier to aim for a combination of diversity and scale advantages at an international level. Both aspects call for a certain degree of coordination and co-operation between countries, as well as the creation of an international extended level playing field as outlined in Section 3.3.

In the Netherlands and many other countries, the dominant electricity generation method nowadays is power plants running on natural gas and coal, supplemented by the co-generation of heat and power. Reasoning from transition policy, it would be logical to create different types of systems that involve a variety of energy carriers, energy sources and technologies. The question, however, is whether such variation can be handled at the Dutch level or should be elevated to a higher political level. The European Union (EU) evidently can play an important role here, not only because of the realized political and economic integration, but also because it harbours a large variety of energy sources and technologies and because energy distribution is becoming increasingly international in scope.

The goal of diversity at the international level can benefit from consultation and coordination between countries; for example, within the context of the EU or the United Nations. In this way we can establish which parties focus on specific sustainable resources or energy carriers that take favourable selection factors into account, such as climate (total hours of sunlight and intensity of solar energy), geography and resources (rivers and wind). This can provide a certain balance between scale and diversity. In other words, the spatial heterogeneity of constant and restricted environmental conditions (notably geographical and climate-related circumstances) can thus be adequately translated into a pattern of spatial diversity of technology and organization of the supply of energy.

Diversity of initiatives can, on the other hand, also come about through a lack of coordination or better decentralization. This is especially relevant at the sub-national level. If the decentralization of energy supply is taken seriously, municipalities and provinces might, more than is now the case, take initiatives alongside those of the national government. This could result in a great diversity of initiatives, with the most successful ones eventually surviving and spreading. It is worthwhile considering, for example, that several energy transitions in the past (coal and electricity) started at the local or provincial level. To implement such an approach, current barriers in terms of hierarchy and spatial planning may need to be levelled.

3.8   CONCLUSION

Policy and institutions appear different from the viewpoint of evolutionary economics than from that of traditional economics. A key difference is the emphasis on diversity as opposed to efficiency. Within the evolutionary approach a diversity of options is regarded as essential for adapting to changing circumstances and preferences, through selection and new innovations. Having said this, policy-setting must not be directed at predetermined results, but at improving the way in which variety selection and innovation processes operate. This relates both to generic and to specific policy. In the first place, transitions require a shift in policy, with a focus on learning processes aimed at system innovation.

Diversity is the key to an environmental policy based on evolution. But the creation and maintenance of diversity implies waste: of time, money and intellectual capacity. Without waste, however, few innovations would occur. Wastage must therefore be fostered. Fisher's theorem is once more stated, as it is particularly relevant in this regard: 'The greater the genetic variability upon which selection for fitness may act, the greater the expected improvement in fitness' (Fisher, 1930). In other words: there can be no major innovations and no transition to sustainable development without wastage and even without dead ends. Unnecessarily tight planning and premature selection of winners lead only to short-term gains and can turn out to be unwise from a long-term perspective.

Pursuing this line of thinking, there is no such thing from the viewpoint of evolutionary economics as an optimal policy. Economic agents behave according to bounded rationality, which expresses itself in the form of habits and routines, imitation of others and use of a limited time horizon. The government plays multiple roles in environmental policy, energy policy and innovation policy aimed at sustainable development. Incorporating the costs of external effects is necessary, but much more is needed. Government should stimulate and guide the exchange of information between players, so that a collective learning process takes place. Subsidies may be needed to initiate major innovations and in particular to avoid or rectify situations of techno-logical lock-in. Subsidies and government tendering policy can stimulate the emergence of niche markets, which accelerate learning curves and increasing returns to scale of innovative technologies, products and processes. From an evolutionary perspective, stimulation of niches is a specific case of fostering isolation, which serves as an important evolutionary cause of major innova-tions. This is witnessed by, for example, the critical role of islands and geo-graphical barriers such as mountains and rivers in evolutionary natural history. Governments play a further key role in the support of scientific research into fundamental innovations, since payback periods and uncertain-ties deter private parties from making significant investments here. Scientific research into transitions is a specific issue that warrants attention in scientific research. This might especially try to draw lessons from both unsuccessful and successful transitions in human-economic history. There is still much to be learned in this respect.

The key element in the transition to sustainable development – including sustainable energy provision – is the creation of a level playing field. This comprises a number of features. First, it means regulation of the external costs of environmental pressure (through either command-and-control or market-based instruments), as well as the positive external effects of inno-vation (through subsidies or patents). In addition, there must be a con-scious effort to keep options open, with proper consideration of learning curve effects, path dependence ('historical accidents') and the selection environment. For example, the fact that a specific technology (sustainable or otherwise) is further along the development path (learning curve) than another need not imply that such a winner must be financially and politically backed to the exclusion of others. The basic consideration comes down to keeping options open versus enjoying the increasing returns to scale of a dominant technology and organizational structure. This can be cast in terms of long-term versus short-term net benefits or efficiency, respectively.

Government must not focus on picking winners, but on creating a suitable environment for innovation and selection. The evolutionary notion of selection must therefore not be confused with the idea that government actively selects technological options. Evolutionary selection means that certain alternatives or options perform better than others and thus survive and replicate – which may include the common socio-economic mechanism of imitation. Some dimensions of the selection environment, such as physical and geographical factors, can hardly be influenced, whereas others are much easier to adjust in a favourable direction. It makes sense to explore the selection environment carefully in advance before deciding on a specific innovation goal. Only then can a good overview be obtained of the potential focus areas of transition policy.

From a co-evolutionary perspective on transitions it is wise to determine, for intended innovations, which technological complementarities can play a role. These may then be subjected to specific stimulating and coordinating policy. This can be based on taking into account the structure of the economic sector, or the roles of suppliers and customers. An important consideration is that innovations should, as much as possible, reflect a diversity of complementary technologies. After all, if the commercial success of a desirable technological innovation (e.g. the portable computer) strongly or even fully depends on a specific complementary technology (e.g. battery technology), the lack or incomplete development of the latter will act as an unwanted barrier to the former.

At the national level, a balance between diversity and scale may not be easily realized. At the international level, however, it will be generally easier to accomplish due to a larger scale of the combined economic activities and markets. To benefit from international opportunities, coordination is needed through co-operation between countries with regard to the transition policy focus as well as the direction and magnitude of public and private investments in relevant innovation trajectories. Aside from that, the diversity of initiatives – in technology, networks, organization, regulation and institutions – can be stimulated at the local level, ranging from municipalities to provinces. To achieve such spatial diversity of 'sustainability initiatives', various policy barriers may need to be removed.

As for the specific issue of transition to a sustainable energy system, efficiency improvements may hinder the speed of a transition, since they enhance the lock-in of the current system that is based on fossil fuels. It is to be recommended therefore that, in aiming for energy efficiency, priority is given to methods for the generation, transport, conversion and consumption of energy that do not negatively discriminate against renewable energy sources (wind, sun, water). It is tempting to focus entirely on energy conservation of fossil fuels, since much knowledge is readily available and since many opportunities are available because of a currently very high absolute level of energy consumption. However, this can hold back achievement of the long-term energy transition goals, as fossil fuels and related technologies become more efficient and thus both economically and environmentally 'cheaper', but without being able to completely eliminate environmental pressure, to say the least. In other words, improvements in the energy efficiency of fossil fuel combustion may reinforce the existing lock-in. A similar reasoning applies to CHP as well as to clean fossil fuels (CO_2 capture and storage), since these can also lead to reinforcement of the lock-in of fossil fuels. In terms of policy, a proper response would be to create critical technical and institutional preconditions for energy conservation and co-generation, aimed at maximum flexibility in the direction of renewable energy as a primary source. Sustainable energy production (using renewable sources) might also be given an extra stimulus to compensate for extra efforts directed at energy conservation. In this way, the negative effect of increased conservation and other efforts on the transition to renewable, sustainable energy can be offset.

The complexity of the energy system suggests that, when deliberating about a sustainable system and a transition to it, a variety of scenarios must be tried out. A systematic approach would begin by making a list of all possible combination of elements, such as the extent of centralization (or decentralization), the type of energy source(s) and carrier(s) involved, the specific technologies employed and the degree of system integration (notably of electricity generation and transport system). This approach would involve scenarios that focus, for example, on solar energy, wind energy, a hydrogen economy, nuclear fusion, nuclear fission, energy conservation or clean fossil fuels. Working backwards from these scenarios (backcasting), potential transition processes might be identified, which in turn could serve to determine effective selection environments as well as required policy measures and institutional arrangements.

Given the complexity of energy systems and their interaction with the rest of the economy and society, as well as the range of criteria and objectives (efficiency over a certain time period, environmental sustainability, security and reliability of energy and electricity supply, etc.), it is crucial to effectively bundle relevant knowledge. This not only provides a test of whether specific options are realistic, but can also lead to innovative combinations of current ideas and insights that had not previously been synthesized. A network of transition experts may be a concrete way to bundle currently diffuse knowledge. And a transition panel might play a role in bundling relevant expertise, as well as evaluating the innovative character and realism of alternative options. Such a panel could keep a constant eye on whether diversity and efficiency are in balance, making sure that neither of these plays too dominant a role, at least until a transition has been realized.

Three general comments can be made to close this chapter. First, transition management is the sum of three specific approaches; namely, environmental regulation, unlocking policy and fostering of innovations. A credible transition policy seeks the integration of the these three specific approaches.

Second, the fundamental consideration from the perspective of evolutionary economics is to find a proper balance between efficiency and diversity; in other words, between short-term costs and the long-term benefits of diversification. This will often entail expert judgements, public opinion or intuition, since there are as yet no practicable guidelines to deal with this trade-off. The development of a good method for making such an assessment is an important task for further research. Most probably, under extreme uncertainty more options need to be kept open than under limited uncertainty. Evolutionary modelling and real options theory, or even a combination of these approaches, might serve as useful methods to explore this issue.

A third general comment is that evolutionary economics describes the economic system in principle as an autonomous and aimless process. Transition management adds normative elements to this system through instruments and institutions aimed at stimulating sustainable innovations. As a result of this, evolution changes from an autonomous and aimless process into a partially managed and steered process on the basis of politically determined goals and boundaries. The concept of aimless evolution is comparable, by the way, with the concept in traditional economics of inherently aimless functioning of the market. Just as policy measures to correct market failures ensure, in such a context, that the operation of the market will be at the service of social welfare, so transition management can put the evolutionary changes in an economy at the service of long-term welfare and environmental sustainability.

Tables 3.1 and 3.2 present a summary of the key elements of this chapter. Table 3.1 gives an overview of general policy suggestions specifically related to the six core concepts of evolutionary economics, as identified and discussed in Section 2.4. Next, Table 3.2 provides a synopsis of the evolutionary perspective on transition management.

Table 3.1 Environmental policy and transition management implications arising from evolutionary economics

Evolutionary-economic concept	Effect	Barrier/Opport-unity (B/O)	Desired policy response and examples of instruments	Comments
Population with diversity of agents, strategies or technologies	Diversity (in technology, organizational structure, firm strategies etc.)	O	Nourish and stimulate diversity of options; no fear of wastage as this is short-term thinking; wastage is necessary as it stimulates diversity and indirectly opportunities for innovation and selection; this will lead to fitness improvements and a better system in the long term.	
Innovation	Greater diversity, improvement of desired performance	O	Stimulate major innovations through cross-fertilization and (re)combinations of diverse alternatives; for example, multiple forms of renewable energy alongside each other.	
Selection environment	Less diversity, e.g. through imitation; improvement of performance	O and B	Correct prices, control liberalization (market operation), privatization (networks) and internationalize regulations for sustainability (especially focused on emission of greenhouse gases); stimulate green funds (aimed at wind and solar energy) and stimulate consumers (e.g. offer alternative choices, such as 'green energy').	
Bounded rationality	Limited time horizon (myopia)	B	Extend time horizon via subsidies, soft loans, patent rules, stimulation of niche markets etc.	

Table 3.1 (continued)

Evolutionary-economic concept	Effect	Barrier/Opportunity (B/O)	Desired policy response and examples of instruments	Comments
	Behaviour characterized by habit and imitation	O and B	Influence behaviour: information, education, exemplary behaviour by influential persons.	
Path dependence and lock-in	Increasing returns to scale and lock-in of inferior technologies (from a social welfare angle)	B	Create an *extended level playing field* by internalizing external costs in prices, supporting technologies at the start of the learning curve, equalizing the time horizon, adjusting for increasing returns to scale. Prevent (too) early lock-in of both unattractive and attractive options; maintain a minimum of diversity so as to facilitate flexibility (resilience) and transitions (unlocking) in the future.	Mechanism needed to ensure ongoing diversity; guideline needed to be able to assess the benefits of diversity versus economies of scale; criterion needed to cut off or stimulate options.
Co-evolution	System that is difficult to predict and manage: characterized by non-linear, unexpected and rapid dynamics	B	Understand evolving subsystems and related interactions so that co-evolutionary mutual feedbacks can be determined. If appropriate, make systems independent when feedbacks work adversely, while stimulating positive feedback.	All transitions are characterized by co-evolution, making them complex and difficult to predict and manage.

Table 3.2   Possible policy implications of evolutionary economics with regard to long-term energy supply

Find a proper balance between efficiency and diversity, or between the short-term costs and long-term benefits of diversification. Given uncertainties involved in the impact of diversity on future innovation and selection, however, finding a balance cannot be treated as a traditional question of cost–benefit analysis. More research is needed to resolve this issue. It is likely that a good strategy is: the greater the uncertainty, the more diversity needs to be maintained or stimulated. At the international level it will be easier to find a balance between efficiency and diversity because of the much larger scale. This suggests a need for more co-operation and alignment of policy, innovation programmes and investments between different countries.

Generate a large number of creative future scenarios based on combinations of core parameters:
- Centralized versus distributed.
- Extent of system integration (e.g. transport and electricity generation).
- Energy sources.
- Energy carriers.
- Conversion technologies.

Perform an initial screening of scenarios. Translate remaining scenarios into options with time paths ('backcasting'): generate diversity of options.

Create and maintain an extended level playing field to keep all chosen options open and give them a fair chance:
- Incorporate external effects in prices.
- Abolish subsidies that are not economically justified.
- Undo or prevent lock-in (create niches; tackle market power).
- Correct for (unequal) increasing returns to scale: stimulate promising technologies that are at the beginning of the learning curve; refrain from stimulating those that are relatively far along the learning curve.
- Be careful to invest extensively in options that are cost-effective in the short run (e.g. energy efficiency improvement, combined heat and power and carbon sequestration), but that may be inefficient in the long run as they delay transitions to sustainable (notably renewable) energy sources and systems.
- Stimulate scientific research aimed at fundamental and high-risk innovations.
- Make innovation agreements with large established players.
- Convert ideas of scientists into applied technologies (e.g. stimulate new enterprises, foster communication between academia and companies).
- Ensure diversity of energy carriers, energy sources and centralized/distributed energy supply systems; give local initiatives a chance (foster diversity of urban/regional approaches).

Table 3.2   (continued)

Combine knowledge related to transitions and system innovations (i.e. support transition networks).

Set up a transition panel of independent experts with the aim of:

- Bundling relevant knowledge and thus stimulate the combination of unusual concepts, ideas and insights.
- Assessing innovation opportunities, barriers, and the potential and realism of specific options.

Create a sense of urgency through adequate supply of information and education to consumers/citizens.

Stimulate scientific research into transitions: development and application of methods, competing and complementary theories, testing of hypotheses, modelling complex systems, and learning from historical cases (descriptive and statistical analysis).

4. Evolutionary policy for energy innovations

4.1 INTRODUCTION

This chapter describes in general terms current Dutch government policy with regard to the stimulation of innovations in the energy sector. The following key question serves as our starting point: To what extent are the views underlying Dutch public policies relating to innovations in the energy sector, as reflected in available policy and advisory documents, in line with the insights of evolutionary economics?

Various policy domains are relevant for the stimulation of energy innovations. These include energy policy, environmental policy, climate policy, technology or innovation policy, scientific policy and transition policy. In this chapter we will look in particular at energy, technology and innovation policy. The other areas of policy will be covered only insofar as they are directly relevant for innovations in the energy sector.

The chapter starts with a brief summary of Dutch policy and advisory documents that underlie our analysis of evolutionary principles in energy and innovation policy. The focus is placed on interaction between, and motivation for, energy research policy and innovation policy. Section 4.3 briefly summarizes the history and background of Dutch energy policies, including the policy design for energy innovations. Section 4.4 presents a similar historical overview of Dutch innovation policies. In Section 4.5 we evaluate the Dutch design of energy and innovation policies against the background of the evolutionary concepts from Chapter 2.

4.2 POLICY AND ADVISORY DOCUMENTS

Energy policy goes back decades and can be traced quite clearly since the 1970s. Technology policy is more difficult to trace, because initially it fell under the umbrella of industrial policy. Nowadays, politicians prefer to speak of innovation policy for stimulating innovations and technological developments. Transition policy was introduced in the Netherlands in 2001 in the Fourth National Environmental Policy Plan (NEPP-4) (VROM,

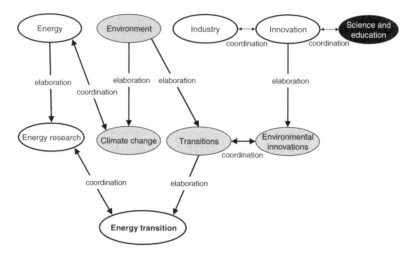

*Figure 4.1   Dutch energy innovation policies embedded in a range of
             different policy fields and coordinated by a number of
             ministries (each ministry carries its own tone: white for
             Economic Affairs, grey for Housing, Spatial Planning and the
             Environment, black for Education, Culture and Science)*

2001). Transition policy is a system approach that has its origins in standard environmental policy, but it goes beyond the ecological sphere. It has
clear overlaps with energy and innovation policy.

The complexity of policy regarding the stimulation of innovations in the
energy sector relates to the commitment of various ministries and related
directorates to the issue. The Ministry of Economic Affairs (abbreviated to
EZ in Dutch) is largely responsible for energy policy and overall technology
and innovation policy. Environmental policy is coordinated by the Ministry
of Housing, Spatial Planning and the Environment (abbreviated to VROM
in Dutch). This ministry thus also plays an important role in the stimulation
of innovation in the environmental sector. Climate policy links VROM with
EZ, as it involves elements of environmental policy as well as energy policy.
Transition policy is essentially the long-term strategy of environmental
policy, although the scope of transition policy may be broader than most
other policy areas (even encompassing them). Transition policy is coordinated by VROM, but the coordination of its separate components (agriculture, energy, biodiversity and mobility) is distributed over various ministries,
with EZ managing energy transition. General scientific policy is laid out by
the Ministry of Education, Culture and Science. Figure 4.1 shows the overall
relationship between the policy domains that directly influence energy

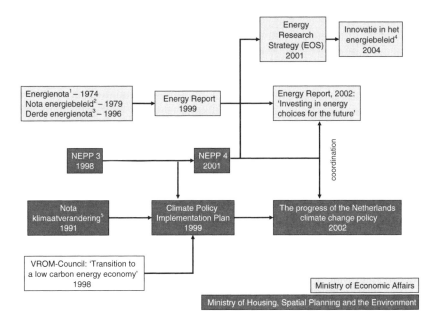

Notes: *Translations of Dutch report titles:*
1. 'Energy Report';
2. 'Energy Policy memo';
3. 'Third Energy Report';
4. 'Innovation in Energy Policy';
5. 'Climate Change memo'.

*Figure 4.2   The relationships between the policy documents of the various
departments in connection with energy and climate policies*

research and energy innovations. Figures 4.2 and 4.3 show the relationship
between the various memoranda and advisory reports that have played a key
role in the development of current policy. Obviously, many other policy areas
impact energy research and the application of energy innovations directly or
indirectly, such as education, economic policy, spatial planning and external
safety. These policy areas are not covered in our analysis.

4.3   ENERGY POLICY

4.3.1   An Historical Overview of Dutch Energy Policy

In 1974 the Ministry of Economic Affairs published its first policy document on energy (EZ, 1974), in response to the energy crisis and the report

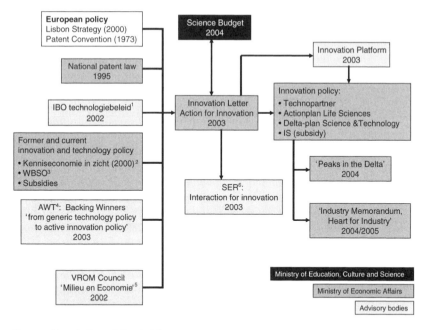

Notes: Translations of Dutch titles:
[1] 'Interdepartmental policy research on technology policy';
[2] 'Knowledge economy in sight';
[3] 'The support of research and development act';
[4] Advisory Council of Science and Technology Policy;
[5] 'Environment and economy';
[6] The Social and Economic Council of the Netherlands.

*Figure 4.3   The relationships between memoranda, advisory reports and
                institutions in connection with the design of innovation policy*

to the Club of Rome (Meadows et al., 1972). This document was followed
by a more strategic outline on energy policy some years later (EZ, 1979).
The main driving force for energy policy, as reflected in these documents,
was the growing energy demand on the one hand and the increasing and
vulnerable dependence on oil from the OPEC countries on the other.
Energy conservation thus received top priority in Dutch energy policy in
the 1970s and 1980s. In addition, diversification of fuel inputs for the production of electricity was considered essential to reduce the dependence on
oil and gas. Nuclear energy was regarded as an excellent alternative to fossil
fuels and in the first half of the 1970s well over half of public energy
research funds went directly towards the development of nuclear energy. In
the years thereafter these budgets were significantly reduced compared with
other energy themes, due to growing social resistance.[2] Investments were

increasingly directed to research on energy conservation and on the development of alternative and more sustainable energy production options (Figure 4.5).

In the 1980s, acidification and climate issues became ever stronger driving forces in energy policy and research into sustainable energy sources. In 1990, EZ published its policy document on energy conservation and the next year VROM published a policy document on climate change (1991), which leaned heavily on the policy document on energy conservation with regard to actions in the field of energy. Increasing the efficiency of energy consumption (i.e. energy conservation) was regarded to be the most important policy measure in the energy sector with regard to the reduction of CO_2. In addition, research and incentive programmes were announced for fuel cells, co-generation, solar boilers and solar cells, biomass and wind energy. The possibility of a regulating energy or CO_2 tax was first investigated at that time, but not yet introduced as a policy measure. In the 'Third Policy Document on Energy' (EZ, 1996), reduction of the dependence on a small number of often unstable countries producing fossil fuels was still the main driving force in energy policy, but the issue of climate change took on an increasingly prominent role in shaping Dutch energy policy. The liberalization of the European energy market has become an increasingly dominant theme in Dutch national energy policy since the late 1990s, especially because of its potential impact on energy supply.

4.3.2 The Underlying Premises of Recent Energy Policy

Liberalization of the energy market and the goal of CO_2 emission reduction as set out in the Kyoto Protocol are the most important starting points for current Dutch energy policy. According to the Climate Policy Implementation Plan, the government hopes to achieve most of the Kyoto goals through energy conservation and sustainable energy (VROM, 1999). Following advice from the VROM Council in 1998, this document speaks of the need to come to a transition to a low-carbon energy economy. The concept of transition is elaborated in NEPP-4:

> Solution of the major environmental problems calls for system innovation, which can be shaped in various ways. In some of these, innovation must be effected through a long-term social transformation process (often involving more than one generation). This transformation involves technological, economic, socio-cultural and institutional changes, which interact and must reinforce each other. (VROM, 2001)

The problem of climate change is one of these persistent environmental problems and a 'transition to an economy based on sustainable energy' is thus regarded as necessary (VROM, 2001). Energy transition policy has been developed further in several policy documents: 'Energy Research Strategy (EOS)' (EZ, 2001), 'Energy Report 2002; investing in energy, choices for the future' (EZ, 2002), 'The progress of the Netherlands climate change policy' (VROM, 2002) and 'Innovation in Energy Policy' (EZ, 2004a). In addition, various recent policy documents and action plans cover the transition to a sustainable energy system.[3]

A consequence of the liberalization of the European energy market is the declining stimulus for the energy sector to conduct long-term strategic research, which is necessary to support the transition to a sustainable energy system. This topic has therefore been elaborated in public policy, specifically designed to stimulate long-term energy research. This is largely attended to in the 'Energy Research Strategy' (EOS) (EZ, 2001).

4.3.3   The Development of Energy Policy to Stimulate Energy Innovations

The stimuli for energy innovations come from two national policy goals, both of which are derived from the overall Kyoto targets. The first goal is to achieve an energy saving of 1.3 per cent per year. The second goal is to generate 10 per cent of total energy consumption by 2020 from sustainable energy sources (VROM, 1999). In the explanatory memorandum to the budget of the Ministry of Economic Affairs for 2004, the operational targets for energy policy are formulated as follows (EZ, 2004a):

1. *Energy efficiency*: improvement by an average of 1.3% per year;
2. *Sustainable energy*: to constitute 10% of total energy consumption by 2020;
3. *Sustainable ('green') electricity*: 9% by 2010 (interim goal: 6% by 2005);
4. *Reduction of CO_2 emissions*: 9.4 megatonne of CO_2 equivalent per year in 2008–2012 (base year 1990).

The strategic energy research policy document EOS distinguishes between long-term and short-term research. With regard to short-term research, the focus lies on regulatory and economic measures, such as tightening standards and influencing the pricing structure. The government sees a role for itself here by providing support, acting as a broker and a means of transferring expertise on behalf of the parties who are involved in short-term energy research and the application of new energy technologies (EZ, 2001). With regard to long-term research, the government sees itself mainly in a 'director's' role, consciously steering and setting out the route to be taken. Two criteria are decisive for the focus of Dutch energy research policy: the international position of the Netherlands in the field of research

	Contribution to sustainable energy economy	No contribution to sustainable energy economy
NL has international leading position	Priority topics	Export of knowledge topics
NL does not have international leading position	Import of knowledge topics	Non-relevant topics

Source: EOS (EZ, 2001).

Figure 4.4 The decision matrix for long-term energy research in the Netherlands (NL)

and its contribution to an economy based on sustainable energy. These criteria are set out in the decision matrix shown in Figure 4.4.

The *priority topics* of energy research are areas in which the Netherlands holds an international leading position as well as areas that contribute to a sustainable energy system. To prevent fragmentation of research efforts, no more than five priority topics are selected. Research areas that are important from an energy perspective, but where the Dutch knowledge position is weak, are flagged as *import themes*. According to EOS, it is essential to uphold a knowledge base for these themes in order to be able actively to import further knowledge and make use of innovations on these themes. Public stimulation of R&D on these themes is not considered to be an appropriate policy. On themes with no contribution to a sustainable energy system but with a leading international position, no active government support for research is provided. Themes with no contribution to a sustainable energy system nor with a leading international position for the Netherlands are not regarded to be relevant for public R&D policy.

To make energy research policy more concrete, EOS recommends a mix of policy instruments that take the distance to market introduction into account. In doing so, the government increasingly utilizes the instruments available in generic innovation policy (see Section 4.4). This leaves room for financial support of the priority themes and for the import themes. This financial support is focused on co-financing (support of research at institutes of knowledge/expertise that are managed and financed mainly by the business community), long-term strategic research, demonstration projects and enhancing the knowledge infrastructure. Various institutions play a role here, from fundamental research units at university level to application units within the industrial sector. Intermediary organizations such as the Energy Research Centre of the Netherlands (ECN) play a supporting role for market players.

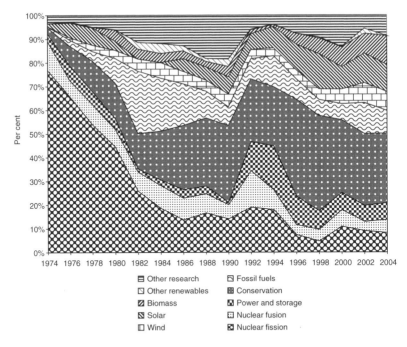

Source: IEA (www.iea.org).

*Figure 4.5   A breakdown of energy research funding in the Netherlands,
1974–2002 (relative shares)*

The Dutch energy research budget for 2001 amounted to approximately
€310 million, of which some €140 million was financed by the national
government. The European Commission contributed approximately
€20 million. The remaining €150 million came from private investors
(EZ, 2001). Over the years, energy research funds have been used for quite
different purposes. In the 1970s most research funds went into nuclear
energy. In the 1980s and 1990s, however, the funds that went into energy
conservation and sustainable energy (such as solar energy, wind energy,
biomass and waste treatment) grew significantly (see Figure 4.5).

Figure 4.5 shows that the diversity of energy research has grown substantially since the 1970s. Nevertheless, EOS has been criticized for being
too narrow, with a focus only on a small number of hopeful energy technologies, and without taking into account learning effects and experiences
(Kets and Schaeffer, 2004). The costs per unit of sustainable technologies
are, however, prone to drop due to (private or public) R&D investments
(learning by investing) and through application of technology (learning by

doing). The critics argue that government focuses solely on R&D and on demonstration projects, failing to take these learning effects into account.

Since the publication of EOS, energy research policy has developed further along the transition route; in particular, where it focuses on the long term. With regard to energy research policy, the Dutch government prefers to collaborate with the business community. The policy maker does not select specific technological options beforehand, but organizes its policy around a cluster of options, the so-called *main routes for a sustainable energy system*, intended to provide a clustering anchor for market players. Five main routes have been selected (EZ, 2004a):

- *Efficient and green gas*: this specifically takes into account the preparations for a hydrogen economy in the future, which could make use of the natural gas infrastructure.
- *Chain efficiency*: conservation of energy and materials across the entire chain.
- *Green raw materials*: biomass to replace fossil fuels, but also as a raw material for the chemical industry.
- *Alternative engine fuels*: to replace oil products.
- *Sustainable electricity*: in particular, biomass and wind, but also options in the built environment.

The intention of this research agenda in energy transition policy is first to explore these main routes further in the coming years and to arrive at increasingly more concrete attractive options from a sustainability perspective. In practice this involves transition experiments, with €35 million already reserved in 2004. The EOS programming and the knowledge infrastructure will be adapted to the transition approach. There is an explicit possibility of later applying the lessons and experiences from energy transition policy in generic energy or innovation policy; as is, for example, conceivable in the application of tax facilities for green investments. Finally, energy research policy explicitly aims to embed Dutch experiences in international networks of co-operation, where foreign partners will be granted the opportunity to make use of the Dutch 'experimental niche' (EZ, 2004a).

4.4 INNOVATION POLICY

4.4.1 An Historical Overview of Dutch Innovation Policy[4]

Until a long way into the 1970s, different departments applied a range of strategies in order to frame Dutch science and technology policy. In general, the American example was followed: science as the engine of progress. This example was based on a decidedly linear concept of technology development. In 1979 the Ministry of Education and Science published a 'Policy document on innovation', which became an important point of departure for innovation policy in the years to come. The role of science and technology changed from 'progress engine' to 'problem solver' (Kern, 2000). The policy focus shifted from the development of innovations to their application and introduction on to the market. This approach showed especially strong tendencies towards development into a sort of industrial policy, in which individual firms received support for the development of specific technologies.

The 1980s saw the bankruptcy of this industry-focused policy due to some massive business failures, such as happened with Rijn–Schelde–Verolme (RSV), a shipbuilding concern. Innovation policy then became increasingly generic, but with special attention to a number of strategic technology fields such as information technology, biotechnology and materials research. Attention shifted away from the development and supply side of technology to the support demand-pull of technologies, with an orientation on articulation of the market and transfer of knowledge. This development appears to be very much in line with a general shift in the focus of innovation policy at that time (cf. Freeman, 1996). By the end of the 1980s the focus of Dutch innovation policy had fully shifted from the development to the application of knowledge. In 1987 the Dekker Committee had also advised the government in this direction: focus on the transfer and use of knowledge for the development of new products so as to enhance the competitive power of Dutch industry, instead of focusing on the development of knowledge. This approach was also criticized, especially by the then Advisory Council for Scientific Policy (RAWB), which highlighted the risk that the Dutch could lag behind in the field of fundamental research (Kern, 2000).

In the 1990s many companies dismantled their R&D units or transferred them to lower-cost countries abroad. In this period the strategic design of innovation policy became increasingly integrated in accordance with three main principles: strengthening of the networks between the public and private sectors, enhancement of the research infrastructure and better social embedment (Kern, 2000). This was the next step in the integration of innovation policy into other policy areas, a trend that continued until the turn of the century. In general, the whole approach to innovation policy became increasingly generic, as exemplified by the establishment of tax schemes such as the Act for the Support of Research and Development (WBSO).

It is interesting to note that the development of environmental technology was still considered to be a very linear process in the policy document 'Technology and environment' (1991). The precise estimate in this document on the potential of environmental technology is also very remarkable:

> Roughly half the emissions of harmful substances can be eliminated through technology that is already in use on a small or large scale. An additional one-sixth can be solved using technology that is currently in a test phase. The rest must come from technologies that are in the research phase or from entirely new technologies. (EZ and VROM, 1991)

Towards the end of the 1990s the linear concept of technology development was fully replaced by a system approach referred to as *cluster policy*. This approach is built around networks of companies and institutions of knowledge, with the development, transfer and use of knowledge as a policy element. Economic instruments focused on the elimination of market imperfections, especially through deregulation and privatization (see Kern, 2000, with reference to Roelandt et al., 1997).

Broadly speaking, a shift can thus be seen from supply-driven innovation policy in the 1970s to a system approach that takes all functions of the innovation system into account (see Figure 4.6). Traditional financial instruments are still very much applied, even though there are many opportunities for system instruments, such as innovation networks, integrated programmes and future-oriented studies (Smits and Kuhlmann, 2004). With the maturing of transition policy, however, the increasing development of such system instruments can be observed.

4.4.2 The Underlying Premises of Recent Innovation Policy

In March 2000 the European heads of government confirmed their support of the Lisbon Strategy, intended to make the EU the most competitive and dynamic knowledge economy in the world by the year 2010 (European Commission, 2000). At the June 2001 summit meeting in Göteborg, sustainable development was introduced as a key component of the Lisbon Strategy: sustainability as opportunity for economic growth.

The Dutch government quickly gave substance to the Lisbon Strategy through a policy document dubbed 'The knowledge economy in sight' (EZ, 2000a). This document aimed to introduce a number of structural reforms, such as accelerating the liberalization of network sectors (e.g. energy and telecom) plus accelerated completion of the internal market for financial services. In 2002 this was followed by an influential interdepartmental study to revisit Dutch innovation policy. This study highlighted the social benefits of knowledge development as justification for a theme-based innovation policy (IBO, 2002).

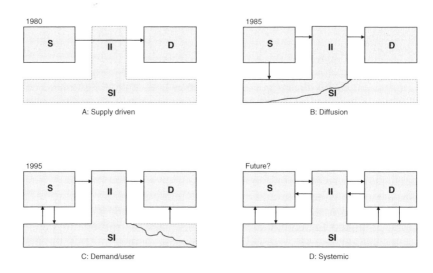

Notes:
D (demand side): users of knowledge services and products.
S (supply side): creation of knowledge involving universities, laboratories, research institutes, etc.
II (intermediary infrastructure): includes innovation centres, policy for technology transfer, etc.
SI (supportive infrastructure): education system, tangible and intangible infrastructure, venture capital, etc.

Source: Smits and Kuhlmann (2004).

Figure 4.6   An overview of the development of the interrelationships in Dutch innovation policy

The problem as well as the policy solutions perceived in many policy and advisory documents in recent years have been viewed more or less unanimously. Policy makers widely regard the further development of knowledge and innovation to be a driver for economic growth and thus a solution for the lack of productivity in the waning Dutch economy (see also AWT, 2003; SER, 2003). In 2003 a new innovation policy was introduced as an implementation of the Lisbon Strategy, starting from this perspective. The new policy document is generally referred to as the 'Innovation Letter' (EZ, 2003b). The number one problem mentioned in the Innovation Letter is that the growth of the Dutch economy is structurally lagging behind due to a lack of innovative strength. The main causes are said to be the complexity of the Dutch innovation system, the shortage of expert workers and the unattractive innovation climate. Furthermore, according to the 'Innovation Letter', there is insufficient *focus* and *mass* in innovation policy; that is, there is too much fragmentation of too little effort and resources. The policy documents in both 2000 and 2003 state that an integrated strategy is needed to tackle the problems. This means that new innovation policy must traverse sectors and technologies, a vision that fits with the general trend seen in other policy fields; in other words, the integration of innovation and technology policy. In this way, innovation policy has finally shifted from a form of supply-driven technology policy to a broader vision of the innovation system as a whole (Quist, 2004).

BOX 4.1   PATENT POLICY

Patent policy is generally designed to protect an inventor against unauthorized use of his or her inventions. In the Netherlands, the protection of inventions and innovations is covered by the Patents Act. This act, originally dating from 1910, was amended in 1995. A patent is the right of the patent owner to the exclusive use of his or her invention. This means that the patent owner may prohibit other persons from using the invention or producing and selling products that utilize the invention. The patent system is an exchange system: in exchange for its protection the invention is made public. As a result of this public access, the scientific community can elaborate on earlier findings, so that technology can develop further. Patents are granted by the Bureau of Industrial Properties (BIE), a government agency that falls under the Ministry of Economic Affairs.

Protection against unauthorized use of an invention makes it worthwhile for an innovator to invest in innovative activities, since a patent system involves a contribution on the part of the user to the inventor. This protection generally expires after a period of about 20 years so that further innovative advancements are not unduly blocked. This period serves as a balance between, on the one hand, private benefits (income) and costs (investment in R&D), and on the other social benefits (the innovation) and costs (higher product price). The protection of inventions and innovations often plays a crucial role in articulating innovative activities, especially in the pharmaceutical and electronics sectors.

At the European level a patent can be applied for at the European Patent Office (EPO), where patents are arranged on the basis of the European Patent Treaty of 1973. This treaty is at the moment being re-negotiated in order to harmonize the various national schemes and to design an acceptable policy with respect to (open source) software.

> At the moment a fierce debate is also going on about the role of intellectual property under the WTO, especially in the TRIPS Treaty (Trade Related Aspects of Intellectual Property Rights). This treaty has been in effect since 1994, but in the current round of WTO negotiations there is a pointed debate about the possibilities of reducing the production costs and increasing the availability of medicines in developing countries.

4.4.3 The Contours and Instruments of Innovation Policy

The legitimacy of government intervention is usually justified by using the argument that market mechanisms do not generate (sufficient) public social benefits. Since the benefits of R&D activities are largely private in terms of knowledge and technology, investment in R&D implies that competing parties may reap at least part of these benefits. Therefore, an incentive for making the investment is lacking and thus *market failure* occurs, which becomes clear through private under-investments for R&D activities. Public support for R&D investments is then justified by the benefits to society at large. Various policy instruments can be directed at market failure; for example, reducing the level of investment risk, shortening the time horizon for profits or increasing private benefits.

The innovation system approach recognizes system failures as well as market failure as an explanation for under-investments in R&D and a lack of market application for research developments. A system failure occurs when the institutional innovation system at large impedes the development, progress and marketing of innovations (VROM Council, 2002). Current innovation policy is based on the concept of the dynamic innovation system, meaning that the development, support, application and marketing of innovations are considered to be interrelated.[5] A dynamic innovation system is a functional unity, integrating the production, transfer and application of knowledge. This approach clearly identifies the opportunities and barriers in the innovation system, allowing policy makers to focus on the failures in the system. An innovation system approach may strengthen the innovation climate, enhance the economic dynamics for innovating companies and take advantage of opportunities in strategic innovation fields (EZ, 2003b). It has been often noted that innovation systems are only as strong as their weakest components or links. It has further been argued that, while the concept of innovation systems is actively applied in strategic policy analyses, its translation into policy instruments is still lagging behind (Kern, 2000; Quist, 2004). One notable exception may be the Innovation Platform, which was installed by analogy to the Finnish example as an instrument to tackle system failure. In Finland this concept played a crucial role in the transformation of an economy that was primarily based on forestry and paper production to one based on advanced technology. In the Dutch case, the main goal is to 'effectively enforce breakthroughs in the Dutch knowledge economy' (Tweede Kamer, 2004), in order to be among the leading countries in Europe in terms of innovative strength. The Innovation Platform demonstrates great trust in technology as the basis for economic growth and progress. The ultimate focus is on fundamental system changes, even though the form that this will take is as yet unclear. The facilitating and transformatory role of the Innovation Platform has thus far not been greatly elaborated.

Current innovation policy is specifically targeted at what is considered to be the weakest link in the Dutch innovation system; namely, the exchange of knowledge between the players involved and the application of innovations in the market. Renewed innovation policy and increased attention to the innovation system as a whole have led to an increased application of generic policy instruments. For example, funding for the generic WBSO regulation, which provides for a tax deduction from the salaries of R&D staff, has been increased considerably. In addition, several former subsidy instruments have been combined in the so-called Innovation Subsidy for Co-operative Projects (IS), for which research consortia may apply, provided that a private partner is included. Such applications are assessed according to four criteria: sustainability, quality, innovation and economic potential. Innovation thus no longer appears to be a goal on its own, but a means to ensure economic growth.[6] Sustainability is also no longer an independent goal of innovation policy but, rather, a minimum criterion that innovations must meet.

It is worth noting that the bankruptcy, both literal and figurative, of the old industrial policy in the 1980s led to prolonged hesitancy on the past of Dutch policy makers to integrate innovation policy with industrial development. Industry policy is very much considered to be state support, a concept that is not only limited by European legislation, but also regarded as a market intrusion from which government should refrain at all times. This 'liberal market' approach has determined Dutch industry policy for at least two decades, as is evident in the following quote from Parliament Member Van Dam (Labour): 'At times the market needs a slight push, as with sustainable energy [. . .]. But it must be no more than a slight push. The government must not take over the investments.'

The Dutch government presented its current view on industrial policy in its recent 'Industry memorandum' (EZ, 2004b). This policy paper focuses on the preconditions for the business climate and overall innovation policy, but it also addresses specific sectors and target groups in the so-called 'key areas approach', which was introduced earlier in the 'Innovation Letter' (EZ, 2003b) and developed further by the Innovation Platform. This platform identifies four key areas: flowers and food, high-tech systems and materials, water and creative industry. Essentially, the 'Industry Memorandum' announces supplementary actions to policy actions from earlier documents on industry, spatial planning, energy and mobility. These supplementary actions include a general reduction in the volume of regulations, a better link between vocational education and the labour market, better interaction between infrastructure for public knowledge and the business community (specifically aimed, for example, at the SME sector and the four key areas), and actions with regard to specific target aspects and a level playing field.

Public funding for applied scientific research is increasingly focused on innovations and market application, showing a linear concept from research to innovations, known as the 'knowledge chain' in 'Science Budget' policy documents (OC&W, 2003). The Science Budget includes funds for the TechnoPartner action programme, which aims to bring together market parties and researchers on technological innovations. In general, the exact sciences programmes aim to stimulate the eventual marketing of technologies. These plans elaborate on instruments such as BioPartner, where the government plays a role in the support of start-up enterprises in innovative activities. On top of this, €100 million (by 2007) is budgeted in the so-called 'smart mix', which is intended first to prevent fragmentation in the innovation system and will be applied to strengthen top-quality research groups at universities and, second, aims to improve the interaction between industry and the Large Technological Institutes in the Netherlands. The Innovation Platform plays a major role in establishing criteria for the implementation of this budget.

Part of Dutch innovation policy is implemented thematically through the BSIK fund, a large subsidy scheme for investments in knowledge infrastructure, specifically on the themes of ICT, spatial planning, sustainable system innovation, nanotechnology and microsystems, and biotechnological breakthroughs in health and nutrition. The fund is fed by the Dutch natural gas revenues. A special method has been developed to evaluate the BSIK projects, applying legitimacy, benefits to overall society and the risk profile of the projects as key criteria (CPB et al., 2003). The focus on efficiency in these criteria can be seen as reflecting a somewhat neoclassical economics oriented perspective in evaluations. The BSIK instrument links up with a thematic system approach based on demand- or problem-oriented solutions.

4.5 EVOLUTIONARY-ECONOMIC ANALYSIS OF ENERGY AND INNOVATION POLICY

The evolutionary economic theory discussed in Chapters 2 and 3 introduces six important evolutionary concepts of technological development and policy: diversity, innovation, selection environment, bounded rationality, path dependence and co-evolution. In this section these concepts will be used to assess current Dutch energy and innovation policy, with the purpose of gaining insight into the extent to which these evolutionary characteristics are reflected in policy. The policy evaluation along the lines of six different concepts explicitly lacks an integrated assessment. The primary aim of this chapter is to show whether the evolutionary principles are – implicitly or explicitly – being picked up, rather than to provide an overall evaluation of Dutch innovation and energy policy.

4.5.1 Diversity

Stimulation of diversity is the key concept of evolutionary economic policy. This implies that the government will generally seek to broaden the number of firms, technologies, products and strategies. The contours of such a policy involve different levels of scale. The government may, for example, strictly focus on sustainable energy and stimulate many different technologies within that particular field. The Advisory Council for Science and Technology (AWT) has stated that generic policy is appropriate for the creation of general conditions for innovation, but not for stimulation of innovations in promising fields. That calls for a more specific and focused policy (AWT, 2003).

The concept of diversity thus relates to the policy framework: generic policy will keep the playing field wide, while specific policy is by definition more focused. The shift from specific to generic strategies and instruments in innovation policy therefore means that diversity can increase. However, this shift to generic policy and the focus on stimulating diversity appears to be mostly confined to technologies, while the importance of diversity in firms, products and strategies is not mentioned in the policy documents. Apart from this, the emphasis on diversity of technologies does not mean that no choices are made at the national level. For example, the strategic energy document EOS clearly opts for sustainable energy, but not for a specific technology within that domain.

The 'Third Policy Document on Energy' puts heavy emphasis on diversity in technologies (EZ, 1996). This technological diversity is deemed necessary in the policy document on 'Implementation of climate policy', since it contributes to achieving the CO_2 reduction targets and to decreasing the vulnerability of energy supply (VROM, 1999). An important starting point is that the government's priority should not be to opt for any specific technology to the exclusion of others, but to set criteria for their selection (EZ, 2001). These criteria, presented in the EOS, are 'international positioning' and 'contribution to sustainability' (see also Section 4.3). The criteria represent in a way a choice for the theme of sustainable energy, so that diversity in energy technology in the broadest sense seems to be limited. Stimulation of technologies in the early phases of the learning curve (such as solar PV, wind energy or nuclear fusion) could, on the contrary, increase diversity among technologies. As the total set of technologies would then be in better balance, the diversity among the various technologies would increase. In the context of energy transition, five main routes have been selected for long-term policy: green resources, alternative fuels, sustainable electricity (wind at sea and biomass), natural gas and energy conservation. This represents a varied set of energy themes, so that diversity is ensured to a certain extent. No specific further preference is expressed within these technological options: 'the transition approach has taught us that it is better not to choose one of these options in advance, but to keep different lines of development open' (EZ, 2004a). The concept of multiple pathways in the transition to a sustainable energy system has also been adopted by the Taskforce Energy Transition, working in 2006. The Taskforce states that trial and error should be accepted and that a diversity of options is necessary in order to be able to reach ambitious sustainability goals. A level of 'waste' and 'error' is explicitly accepted in order to come up with successful pathways. This clearly is consistent with an evolutionary perspective.

The support for solar PV in energy research policy has been declining for the past few years, while biomass and offshore wind energy have been given more policy support in the light of the criterion of the Dutch international position, although support for the actual deployment of offshore wind parks is considered to be low. The Netherlands also has a good knowledge base for solar PV, but the potential contribution of PV to sustainable energy targets is considered to be low (EZ, 2001). The reduced attention paid to solar PV in the policy documents does not, however, correlate with the research funds for solar energy. Total energy R&D in the Netherlands (private and public) for the various forms of solar energy has in fact risen slightly, to just above 10 per cent (see Figure 4.5). Although public research into fuel cells appears to be declining, various firms increasingly engage in activities on this theme. Public policy prefers to see alternative fuels and biofuels as a likely option for the short and medium term, as these options tie in better with the current system of energy production and distribution. The road for radical alternatives (in particular, the hydrogen economy) is not entirely cut off in public policy support, but neither is it fully stimulated. At the macro level, system diversity is thus not fully included in energy research policy.

Conclusion on diversity

To summarize, the importance of maintaining diversity is acknowledged in policy documents, but in practice this applies mainly to diversity in technologies and much less to diversity in companies, products and strategies.

4.5.2 Innovation

Many of the policy and advisory documents mention the lack of innovative strength in Dutch companies as a major problem. Table 2.1 mentions a number of characteristics of the concept of 'innovative strength', to enable an assessment in terms of evolutionary economics. Some of these characteristics (co-operation, venture capital and education) are frequently mentioned in government analyses, while other characteristics (serendipity, isolation, future visions and niche markets) hardly get any attention at all.

Combination, cross-fertilization and serendipity

In the policy documents innovation is generally regarded to be 'the successful introduction onto the market of new, improved or more competitive products, processes, services or organisation forms' (AWT, 2004, p. 13). While the discovery of something new – the invention – lies at the base of 'the introduction onto the market', in this definition it is no longer the focus of policy. Creating combinations, cross-fertilization and serendipity are concepts that specifically relate to the invention, but in the policy documents these concepts are hardly mentioned.

Strictly speaking, combination, cross-fertilization and serendipity do not need to be restricted to inventions, since cross-fertilization and coincidence often also play a significant role in successful marketing strategies (consider, for example, the 3M yellow Post-It® Notes; see Box 4.2). Nevertheless, the concepts of combination, cross-fertilization and serendipity are hardly referred to in the policy documents. The general emphasis is on 'focus' and 'concentration', 'valorisation' (the conversion of research results into economic value), and stimulating competition. Room for experimentation is thus effectively limited.

BOX 4.2 COINCIDENCE AND CROSS-FERTILIZATION AT 3M

The story behind the yellow Post-It® Notes is a classic in three chapters. First, the glue was discovered, resulting from a failed experiment by Spence Silver, performed during the 15 per cent of time that he was allowed by 3M to try out new ideas. The glue hardly dried and Silver therefore used it to stick scraps of paper for notes on the office bulletin board, so that they could be taken off easily after a few days.

In the second chapter, Arthur Fry, Silver's colleague, appears on the scene. He was a choirmaster and he came up with the idea of putting the glue on the paper scraps instead of on the background, such as the bulletin board. This allowed him easily to trace the hymns in the hymnbook during a service, as they were marked by the paper scraps. This was essentially the invention of the Post-It paper.

In the third chapter, marketing plays a leading role. Even though the Post-It® Notes became quite popular within 3M, they hardly sold outside the company. Two marketing employees then distributed the Notes free of charge to a large number of companies, thus effectively creating a dramatic increase in demand, which ultimately resulted in a marketing success story.

See also www.uh.edu/engines/epi726.htm.

In energy transition policy, at first sight there appears to be room for experimentation. Remarkably, however, the opportunity to really experiment is limited by the objective of these experiments:

> The experiments are thus intended to demonstrate something, firstly, of course for the participants themselves, but also for the general public. The purpose of all this is, after all, that the approach, details and results of this first set of experiments create such enthusiasm that new parties, who may still be hesitant, will be challenged to pursue the sustainable route (EZ, 2004a, p. 23)

This objective raises the question of whether these experiments would provide room for investigating new options and ideas, or whether demonstration projects are involved.

Education
Education and schooling are crucial for the level of innovation. Several policy documents focus on schooling and the training of (process) technicians. Nevertheless, the general level of funding for education and schooling in the Netherlands is among the lowest in Europe (NOWT, 2003). A relatively high percentage of education funds is tagged for higher education, whereas SMEs (Small and Medium-sized Enterprises) also encounter problems in finding trained employees at the secondary level of education. The Industry Letter aims to rectify this situation by announcing supplementary actions that are specifically aimed at secondary vocational education

(EZ, 2004b). In addition, the intention to stimulate lifelong learning is to be elaborated (EZ, 2003b).

The 'Science Budget 2004' document claims to seek elimination of the barriers to the immigration of expert workers – in particular, scientific researchers (OC&W, 2003) – although evaluations are mixed on whether the removal of these barriers has actually taken place. The recent measures to eliminate scholarships for students from outside the EU and for students older than 30 appear to contradict the intentions for lifelong learning and immigrant expert workers.

Isolation

In terms of economy and location, the Netherlands is far from being an isolated country. The Dutch economy has for centuries been very open and focused on the outside world. In terms of language, however, the Netherlands is considered by some to be isolated, and several policy measures are therefore aimed at lowering the language barriers by having Dutch students learn foreign languages from an early age onwards. Isolation as a positive factor for innovative experiments does not play a role in any of the policy documents.

Co-operation

The concept of co-operation receives much attention in Dutch innovation policy. A major challenge that is repeatedly mentioned in innovation and science policy documents is to tackle the 'European paradox'; in other words, that while European research is world top level, the number of commercial patents that originate from this research is relatively low. Stimulating co-operation between companies and research institutes is considered to be an important way of improving the utilization of knowledge. The establishment of the Innovation-oriented Research Programmes (IOP) and the Leading Technology Institutes (TTI) are examples of government policy aimed at co-operation between public knowledge experts and private entrepreneurs. A significant element discussed in the EOS is co-financing, where the government directly participates in the funding of research by Technology Institutes that is initiated by the business community.

In energy transition policy much attention is also paid to the concept of co-operation. One of the roles that the government sees for itself in the energy transition is that of a broker, bringing various stakeholders together; for example, market players, institutes of knowledge and public policy institutions. In addition, the 'Science Budget 2004' document also declares a wish to enhance co-operation between universities, institutes of knowledge and the business community. Co-operation within the scientific

community is intended to create the critical mass that is considered necessary for top-quality research. It is not clear how this relates to raising the
level of competition between research groups, which according to the policy
documents is also necessary.

Venture capital
The availability of venture capital is closely related to the bounded rationality of companies and the government (see also Section 4.5.4). In several
advisory documents, the limited availability of venture or risk capital is
considered to be a major limiting factor for the innovative strength of
Dutch industry. The R&D activities of Dutch companies in general lag
behind those of companies abroad (NOWT, 2003). The liberalization of
the energy market appears to have made energy companies less willing to
conduct long-term energy research (EZ, 2002). In 2001 the Dutch government spent approximately €140 million on energy research, while private
parties invested about €150 million. The EU spent approximately €20
million on energy research in the Netherlands (EZ, 2001). The Dutch government is trying in various ways to increase the level of research funding:
first, through direct public funding of research and, second, by making
private research investments more attractive; for example, through the co-
financing regulation for energy transition or within the generic WBSO
regulation.

Niche markets
Niche markets may stimulate innovative activities, but this notion hardly
receives attention in the policy and advisory documents discussed. In a
more implicit way, the stimulation of 'green energy' and the low tax rate for
green funds can be interpreted to constitute niche market development;
namely, as a niche market in the fields of energy supply and financial services. Targeted regulations might also contribute to stimulation as well as
to protection of a niche market by providing advantages to some innovations over competing non-innovative options. On the other hand, regulation could also inhibit the development of niche markets; for example, by
not allowing for market experiments. In energy and innovation policy,
however, regulations are hardly specifically aimed at the creation of niche
markets.

Future visions
In Dutch energy and innovation policy considerable attention is paid to
future visions and outlooks (EZ, 2000b, 2004a). Projects related to innovation policy and energy transition options nearly always start out with long-
term explorations such as outlook studies and scenarios. Policy makers

often consciously use the various roles that future visions can play. For example, scenario studies are often used to identify the 'policy gap' between policy objectives on the one hand and the expected results of developments and implemented policies on the other. Future perspectives are also used to investigate different policy options in a broader context ('backcasting'). Finally, future visions are regularly applied as a means of communication and to achieve a consensus between different stakeholders about the course to be followed. The case studies in Chapter 5 also provide clear evidence of the important role of future visions in policy aimed at stimulating energy innovations.

Conclusions on the concept of innovation
Innovation is a key concept in many policy documents, but from an evolutionary perspective the aspects that are essential for the increase of innovative strength are rather biased. The basic innovative concepts of *co-operation* and *future visions* score highly in evolutionary terms. *Schooling and education* and *venture capital* are mentioned in the documents, but their implementation in terms of concrete policy is rather meagre. Aspects such as *combination*, *cross-fertilization*, *serendipity*, *isolation* and *niche markets* hardly get any attention at all in policy documents, and in some cases, policy measures even impact these aspects in a negative way.

4.5.3 The Selection Environment

The selection environment for innovations is determined by the physics, technology and geography, the business, market and institutional features, and the public policy and specific conditions that affect R&D (see Table 2.1). Policy and advisory documents do not generally mention the physical and technological limitations to innovative activities. Geographical features are mainly addressed indirectly; for example, through wind energy being regarded as a serious option in a windy country. The choice of the Rotterdam harbour area as an experimentation ground for some options in energy transition may have been driven by a combination of geographical and business features, even though this is not explicitly mentioned in any of the documents.

The market
Policy often refers to market conditions, indicating that an efficient market will more or less automatically create the right selection environment for sustainable innovation. Liberalization of the energy market and the impact of oil prices receive a lot of attention in energy innovation policy. The liberalization of the energy market has the effect that public authorities

cannot steer things as directly as in the past: 'from field player to coach' (EZ, 2001). Liberalization also means that energy companies will focus more on short-term developments and less on long-term research (see also Section 4.5.4). Nevertheless, liberalization of the energy market is generally viewed as a positive development, since it is expected to lead to greater economic efficiency of energy supply. In a liberalized market the oil price constitutes a dominant factor in the decision on whether or not to invest in sustainable energy technologies. With a low oil price (or, maybe even more importantly, an unstable oil price) the readiness to invest in sustainable alternatives will be low. Therefore '. . . current energy policy focuses on meeting, as cost-effectively as possible, the goals for 2010 in terms of CO_2 emission reduction, energy conservation and sustainable energy' (EZ, 2004a, p. 6). A limit of the equivalent of €68 per tonne of CO_2 has been set as a criterion for the choice of measures to reduce CO_2 emissions (VROM, 1999). The policy document 'Innovation in energy policy', states that:

> Generic policy aims to incorporate (i.e. to internalize) the external costs of energy consumption [. . .] in the price of energy as much as possible. This is done through tax measures and the use of a European system of CO_2 emission trading. When the social costs are charged to those that cause them, this will yield the right conditions for the market to do its work. A new market-based selection mechanism for innovations thus arises. Public authorities do not need to make any choices here, since the winners will present themselves. (EZ, 2004a, p. 12)

Institutions and public policy
Public authorities pay much attention to the market as a determinant for the selection environment. Government policy is highly focused on eliminating market failures. It does so by publicly financing long-term research and occasionally by including (marginal) external costs in prices; for example, through subsidies or tax measures. Other factors that could determine the selection environment are environmental regulation and institutional investment. Stringent environmental regulations are not mentioned in any of the policy documents as an option to either stimulate or elicit innovations. The role that government sees for itself in energy transition is that of a trusting broker, partner and leader (EZ, 2004a). As for trust and leadership, emphasis is put on being able to guarantee experimentation facilities in the long term. On themes where the government sees itself as (or is regarded to be) a problem holder – for example, with regard to securing energy supply or to setting environmental goals – an active guiding role on the part of government is generally accepted. A clear perspective on the role of public authorities in setting and enforcing strict boundary conditions, thereby steering the selection environment, is absent from the policy documents. Rather, the regulatory role of government is regarded to be a

negative factor for industry, as the perceived ever-increasing regulatory pressure is thought to inhibit industrial activities (EZ, 2004b). In particular, the Industry Letter mentions the high administration costs, lack of uniformity in the interpretation of European rules in the Member States, prolonged uncertainty about permits, inconsistency in regulations and problems of enforcement. Strict environmental policy is generally regarded to have a negative impact on the international competitive position of electricity producers (see also 'Level playing field' in Section 4.5.5).

An interesting option for public authorities to influence the selection environment is through green procurement. The EU issued a number of communications in 2003 within the framework of the Sixth Environmental Action Plan that highlight the importance of such procurements: 'On Integrated Product Policy', 'Towards a Thematic Strategy on the sustainable use of natural resources' and 'Towards a Thematic Strategy on the prevention and recycling of waste'. If for example, according to the European Commission, all European governments were to procure green electricity, this would result in a saving of 60 million tonnes of CO_2, constituting 18 per cent of the EU's commitments under the Kyoto Protocol. With the publication of 'Buying green, a handbook on environmental public procurement', the Commission seeks to support Member States to align their green procurement activities with requirements of other EU directives on public procurement (European Commission, 2004b). In Dutch policy documents, however, the opportunities for green procurement are not mentioned.

Conclusion on selection environment
Selection environment as a concept scores low in current public policy. Somewhat in line with a neoclassical economics perspective, the general leading principle in innovation policy is that 'the market should choose'. Almost by definition this implies that public authorities have little impact on the selection environment. After all, 'government action disrupts the operation of the market'. The present Dutch government is devoting considerable effort to the removal of innovation barriers through the elimination of regulations and the administrative burden. The drawback to this perspective, however, is that opportunities to positively impact the selection environment for innovations that benefit the general interest are often missed.

4.5.4   Bounded Rationality

The time horizon
In its policy the Dutch government takes into account the limited horizon of companies, caused by their bounded rationality: 'since research by the business community concentrates on the short term, the government shifts

the focus to the long term' (EZ, 2001). The 'Science Budget 2004' document and the 'Innovation Letter', plus related advisory documents, extensively discuss the lack of long-term strategies on the part of industry, and to the consequences for long-term research and the related funding (see Section 4.5.2). The remedy in order to stretch the short time horizon in the business community usually focuses on public (co-)funding and initiating research in public institutions. Recently, Members of Parliament requested that the establishment of a guarantee fund for high-risk investments should be investigated, based on the argument that a lack of sufficient venture capital in the Dutch equity market via the regular banks and investment companies is felt, especially when amounts up to €5 million are involved. With regard to this subject it would be interesting to investigate the effect of the Dutch tax on imputed return on investments on the level of risk acceptance by industrial companies. By assuming a standard return on equity investments, equity providers might be less inclined to invest in technologies (or in their development) that only yield a return in the long term. Unfortunately, to our knowledge no empirical research appears to have been conducted into the effect of this recent tax change on the risk acceptance by companies and their equity providers.

In strategic policy documents, government authorities generally take a broader perspective on energy research. Scenarios with a time horizon of 20–50 years are not uncommon. Nevertheless, when implementing energy research policy, again a short time horizon often prevails. For example, in climate policy achieving the 2010 Kyoto targets is top priority, sometimes conflicting with the realization of long-term goals.

Routines
Whereas public authorities regard firms as quite rational in many respects in their management and search for innovations, they themselves often display a surprising sense of routine in their strategies and behaviour. This is most clear in the emphasis on market-driven instruments and pricing when deliberating the choice of policy instruments. Such instruments assume that companies and consumers will react rationally to financial incentives. Another example of bounded rationality at the level of public decision making is the focus on economic growth as a goal instead of as a means of achieving greater welfare. The 'Innovation Letter' introduces knowledge as a central competitive factor in achieving economic growth. On the other hand, however, this letter states that wage restraints are necessary to achieve a competitive cost level (EZ, 2003b, part 1, p. 9), implicitly disagreeing with the economic view that wage restraints will lead to low-level jobs instead of the desired high-quality innovative labour (cf. Kleinknecht et al., 2002). This assertion is still being fiercely debated, since

higher wages could also lead to higher unemployment, after which wage levels will drop again (Huizinga and Broer, 2004). Generally, a focus on labour costs may be positioned within a more traditional economic policy approach (Boschma et al., 2002).

Imitation

The various policy documents show a strong awareness of the relevance of imitation. For many years much attention has been paid to demonstration projects in relation to energy policy. Demonstration of new technologies is also an important item in innovation policy. This is partly elaborated by stimulating the first movers in the hope that they will serve as a model for others. The Ministry of Economic Affairs is well aware that some of the SMEs will never get involved in innovation projects, since it is beyond their core business. Innovation policy has therefore shifted in recent years from encouraging the laggards to supporting the frontrunners, implying an increasing focus on the firms that are actively involved in innovation. This shift involves the creation of breeding grounds for exchange of innovations and creative ideas. This is illustrated by policy initiatives such as TechnoPartner.

In energy transition policy there is a considerable focus on experiments. The purpose of these transition experiments is to have 'the approach, details and results [. . .] which create such enthusiasm that new parties, who may still be hesitant, will be challenged to pursue the sustainable route' (EZ, 2004a). As a means of reinforcing the aspect of exchange, entitlement for subsidy requires an experiment to be led by a market player, in order to be able to show the market opportunities to other market players. Whether successful experiments will be an example to be shown to competitors remains to be seen, however.

The Innovation Platform, which was set up recently, is a good example of imitation by the government itself, as it has adopted elements from the successful Finnish innovation model.

Conclusion on bounded rationality

To summarize, we can state that public policy scores well in terms of awareness of the aspect of bounded rationality; in particular, with regard to the time horizon of industrial firms and imitation behaviour. On the other hand, public authorities themselves often act routinely by considering and applying only a limited number of policy instruments.

4.5.5   Path Dependence and Lock-in

Path dependence often leads to lock-in of technologies and organizational structures: when a system becomes dominant, its leading position is difficult to overcome. Policy making that reflects awareness of path dependence takes routines and interdependence of different technological systems into account as crucial elements. For example, the Advisory Council for Science and Technology advises a link-up with already existing 'innovative hotspots' in order for a positive lock-in for companies to occur (AWT, 2003). A key question is whether selection of technologies will or will not lead to path dependence and lock-in. In our approach we see path dependence as a restriction of the spectrum of selection and innovation outcomes. Policy that is based on evolutionary economics should, in principle, concentrate on strengthening the economic dynamism for the long term (Boschma et al., 2002, p. 179), rather than limiting choices in the short term. Similarly, the shift from specific technology policy to a more generic innovation policy is in line with evolutionary economics since long-term dynamism is stimulated.

The first policy document on environment and technology (EZ and VROM, 1991) linked up closely with the existing structures at the time, especially given the document's heavy emphasis on 'end-of-pipe' and process technologies that can be incorporated into the regular production process. Current transition policy, on the other hand, is directed at system changes, so that changes are no longer restricted to technologies. This idea is strongly impacted by the awareness of path dependence:

> Continuation or intensification of current policy will not lead to adequate solutions since that would ignore the barriers for sustainable solutions. These barriers basically represent system errors in the current social order, in particular, the economic system and presently functioning institutions. Only through system innovations can the barriers be removed and will it be possible to arrive at a real solution for the great environmental issues. (VROM, 2001, p. 65)

Both energy and innovation policy demonstrate awareness of the concept of lock-in through the wish to link up with the relative strengths of the Dutch economy. This may, however, conflict with transition policy, which puts greater emphasis on the changes considered necessary in the energy system, rather than aligning with the relative strengths in the existing system.

Policy will thus need to focus on scaling and learning effects, leading to the increased diffusion of technologies and eventually to the formation of an alternative technological regime. However, the intention to that effect is hardly evident in the policy documents. For example, during a recent parliamentary debate about subsidies for green energy production, the Minister of Economic Affairs argued that: 'Volume alone does not lead to innovation'. There is hardly any reference to the Schumpeterian concept of creative destruction; in other words, the renewal of sectors and industries that may no longer be competitive. From the early 1980s onwards, such industrial policy has not been applied in the Netherlands because of the notion that, from the perspective of the European free market, state support was not appropriate. This has resulted in a blind spot to the potential benefits of a strategic vision on industrial development (see also Section 4.4.3). The Finnish innovation platform is a good example of how a coherent industrial and innovation policy might be designed, since in this strategy the old Finnish industries were replaced by new high-technology sectors. In the United Kingdom, the renewable energy sector is increasingly incorporated within strategic industrial policy, with a focus not only on sustainability and meeting international environmental targets, but also on stimulating new technologies in the long term and making a contribution towards the revival of rural communities (Strachan et al., 2006). These examples offer plenty of justification for incorporating arguments from strategic industrial policy into the spheres of energy innovation research. The establishment of the Innovation Platform in the Netherlands also aligns with these evolutionary economic arguments.

The level playing field
The various policy documents are highly geared to the liberalization of the energy market. A central theme is the creation of a level playing field, implying that different technologies should be able to compete with each other on the basis of equal starting positions: 'the economic potential must stay as close as possible to the technological potential' (VROM, 1999). In evolutionary practice this would imply that technologies at the beginning of the learning curve get an extra push to compensate for the higher investments that are especially associated with sustainable technologies. That involves, as it were, a catch-up of investments that were made earlier for standard technologies and that now function as sunk costs. The subsidy of additional costs for sustainable technologies enables the sunk costs already invested in standard technologies to be overcome, allowing for the creation of a level playing field, in which the new systems can compete with the existing system on a more or less equal basis (EZ, 2004a).

In Dutch energy policy, however, the concept of a level playing field is applied in competition between countries rather than between technologies. It thus focuses on the creation of a good competitive position with regard to foreign energy producers. According to the 'Energy Report 2002', for example, differences in tax and environmental policy among the EU Member States means that we cannot yet speak of a level playing field at this time (EZ, 2002). The 'Industry Letter' refers to the lack of a level playing field for a number of sectors. From this perspective, the 'Industry Letter' announces actions related to the energy price for industrial bulk consumption, natural gas production, the airlines cluster, the aerospace industry and the maritime sector, all intended to achieve a level playing field (EZ, 2004b). The creation of a level playing field is sometimes used to justify the application of specific policy instruments. The level playing field often correlates with the installed power (output capacity) of an energy production technology: greater installed power allows for learning-by-doing, thus lowering the position on the learning curve. Energy prices are also important for the formation of the selection environment: a high oil price, for example, will make it more attractive to invest in sustainable energy alternatives.[7]

Conclusion on path dependence and lock-in
The notions of irreversibility, lock-in and the level playing field have found their way into Dutch energy and research policy. From an evolutionary economics perspective, however, the impact of these aspects on concrete policy appears to be somewhat biased. With regard to lock-in, the discussion is mostly about the fundamental dilemma between the creation of sufficient critical mass and the maintenance of diversity. Prevention of lock-in thus appears to be realized mostly through postponing selection, rather than by conscious stimulation of flexible options. In energy policy this links up with the choice between high-volume versus low-volume production. The treatment of the level playing field does not entirely align with the evolutionary perspective, as it is strictly viewed as the creation of equal competitive positions between producers. The playing field involving technologies that are in different phases of the learning curve has not received any attention so far.

4.5.6   Co-evolution

Co-evolution refers to the link between different technologies or technological systems: developments in one particular system influence developments in another system through their interlinkages and especially their complementary relationships. This principle implies that the learning curve of one technology can influence that of other technologies. It can therefore make sense to invest in technology A so that technology B can benefit from it, although the direction of the technological development is of course hard to predict. It is thus quite conceivable that certain technological developments correlate with developments in other fields of technology. The various policy documents show hardly any cross-referencing. For example, the energy policy documents do not mention the bottlenecks associated with complements that apply specifically to innovation policy, such as the lack of expert workers. The 'Energy Research Strategy' (EOS) document refers to the influence of other fields of research, such as the life sciences, nanotechnology and ICT (EZ, 2001). 'Innovation in energy policy' (2004) gives a concrete example, which asserts that the development of an economy operating on hydrogen can benefit significantly from the already existing infrastructure for natural gas. It is also pointed out that stimulation of sustainable energy systems will lead to competition for the standard systems, giving them an incentive to optimize and raise their environmental efficiency (EZ, 2004a).

Conclusion on co-evolution

From a co-evolutionary perspective, there is as yet limited coherence between energy policy and innovation policy. The different policy areas may well differ in their problem analysis and their approaches to a solution. Furthermore, complementary aspects of different policy fields seem to receive relatively little attention. However, there is an increasing tendency to focus on energy innovations as a pathway in transition policy, thus showing the importance of innovations for sustainable energy systems.

4.6 CONCLUSION

The technical side of innovations has received much attention in Dutch policy documents on energy innovations. In comparison, social and institutional aspects, such as the business environment, products and strategies, have been neglected to some extent. In energy transition policy, however, these aspects are getting increasing coverage. In general, it can be concluded that the current policy documents that are relevant in the field of energy innovations reflect a remarkable amount of thinking from evolutionary economics. Many of the notions listed in Table 2.1 are taken into account in the various documents considered.

In particular, those aspects in evolutionary economics are addressed that do not lead to tensions or contradictions with a traditional, neoclassical-economic focus on efficiency. For example, much attention is paid to co-operation, schooling, scenarios and demonstration projects. On the other hand, the creation of conditions that might stimulate cross-fertilization and serendipity (experimentation, trial and error) seem not to be regarded as a government task. This also applies to the stimulation of technologies that are still at an early stage of the learning curve. Similarly, aspects of evolutionary economics that could lead to conflicts with European regulations on liberalized markets, such as the creation or protection of niche markets, are hardly encountered in the policy documents nor attempted in policy practice. Finally, the policy documents demonstrate that while public authorities regularly take into account the bounded rationality of entrepreneurs (in particular, the limited time horizon), they do not do so in the design of environmental regulation.

Table 4.1 presents our interpretation of how current policy aimed at energy innovations performs in terms of the insights from evolutionary economics that are listed in Table 2.1.

Table 4.1   The scores for policy and advisory documents covered in this chapter with regard to aspects of evolutionary economics

Diversity	0
● Companies (type, size)	−
● Techniques (production)	+
● Products (characteristics)	−
● Strategies (sales, R&D)	−
Innovation	0
● Combination/cross-fertilization	−
● Serendipity	−
● Schooling	0
● Isolation (physical, economic)	−
● Co-operation	+ +
● Venture capital	0
● Niche markets	−
● Futures	+ +
Selection environment	−
● Physical aspects (e.g. thermodynamic boundaries)	0
● Technology (technical feasibility, costs)	−
● Geographical features (including soil, water, wind and sun)	0
● Internal company features (organization)	0
● Markets (relative prices, market power)	+
● Institutions and public policy	−
● Specific conditions affecting R&D	0
Bounded rationality	+
● Time horizon	+
● Routines	0
● Imitation	+ +
Path dependence and lock-in	0
● Irreversibility	+
● Increasing returns to scale (economies of scale, imitation, learning effects and positive network externalities)	0
● Lock-in	0
● Level playing field	0

Table 4.1   (continued)

Co-evolution	0
● Subsystems	0
● Negative or positive feedback	0
● Spatial	0

5. Case studies

5.1 INTRODUCTION

This chapter examines, on the basis of three different energy technologies, the role that the evolutionary economic aspects discussed in the previous chapters play in the practical development of a sustainable energy supply. The ultimate issue is whether the public policy as applied with regard to these technologies has been adequate from the viewpoint of evolutionary economics and, if not, what amendments to policy would be justified based on the insights of evolutionary economics.

The three energy technologies examined are fuel cells (Section 5.2), nuclear fusion (Section 5.3) and photovoltaic energy (Section 5.4). The following criteria were applied in selecting these cases:

- The cases should deal with potential elements of an energy *supply* system that must be based on sustainable sources for the long term.
- Within the limited number of cases there should be maximum diversity concerning the characteristics of the energy technology (such as the extent of decentralization and the variety of energy sources and carriers).
- The cases should involve technologies that have already been covered by public policy, to allow for analysis of that policy.

Each case starts with a brief description of the history and current situation with regard to the technology, the actors involved, the applications and markets (including niche markets), the learning curve, policy as pursued and institutional aspects. This is followed by a discussion of expectations on the technology involved for energy supply in the future. Next, the driving forces and barriers that have been relevant for the development and application of the specific technology are discussed on the basis of the aspects listed in Table 2.1. Each section ends with a number of comments about public policy as pursued in the past and to be pursued in the future, based on the assumptions of evolutionary economics. The chapter ends with a concluding observation based on the findings from the three case studies.

5.2 FUEL CELLS

5.2.1 History and Current Situation

Fuel cells

Fuel cells convert energy that is released during a chemical reaction directly into electric energy, just like batteries. The difference from a battery is that, in a fuel cell, the electrodes do not participate in the reaction. A fuel (such as hydrogen) and an oxidant (oxygen) interact with each other, causing the release of energy in the form of electricity (and heat).[8] Fuel cells have a potentially high yield, especially when the heat that is released is also used productively. In addition, no emissions or noise are released during the energy conversion process.

The fuel cell principle was already known in the 19th century. An article by Humphrey Davy about fuel cells was published in 1802 (Schaeffer, 1998) and in 1839 Sir William Grove made the first fuel cell to operate on hydrogen. In 1897 an article by W.A. Jacques appeared, about a cell he had made that converted coal (via hydrogen) into electricity, with a yield of 80 per cent (Sanders, 1972).

Nonetheless, the fuel cell, contrary to the steam engine, the combustion engine, the steam turbine and the gas turbine, has never really broken through in energy production. There have been only occasional successes, also in the 20th century. For a long time, practical applications of fuel cells were found mainly in the aerospace industry.[9]

In the 1960s there was significant interest in fuel cells, but this faded in the early 1970s, due, for example, to the termination of the Apollo programme and the disappointing results in terms of costs per kilowatt. In the United States, the energy crisis of 1973 quickly led to renewed interest, but in Europe and Japan fuel cell activities remained limited. Only since 1980, when a demonstration project of 4.8 MW got off the ground in the USA,[10] did the rest of the world start to show any interest again.

There are several types of fuel cells, most of which are named after their electrolyte. Several characteristics of the main types are mentioned in Table 5.1. Third-generation fuel cells (SOFC and SPFC/PEMFC) are presently considered to be the most promising (see Voermans, 2004).

Fuel cells and hydrogen

The potential role of the fuel cell in the world of energy cannot be considered in isolation from that of hydrogen as an energy carrier. Even though there are fuel cells that are powered by other substances, most attention these days is paid to fuel cells in which hydrogen and oxygen interact to produce electricity.

Table 5.1   Types of fuel cells

Cell type	Fuel	Electrolyte	Operating temperature	Applications	Generation
Phosphoric acid (PAFC)	Hydrogen*	Phosphoric acid	150–200°C	Distributed electricity supply, co-generation	1
Molten carbonate (MCFC)	Hydrogen, carbon monoxide	Carbonate (ions)	± 650°C	Distributed electricity supply, co-generation	2
Solid oxide (SOFC)	Hydrogen, carbon monoxide	Metal oxide	700–1000°C	Distributed electricity supply, co-generation	3
Solid polymer † (SPFC)	Hydrogen*	Polymer	40–80°C	Transport, distributed electricity supply, co-generation, portable electricity	3
Direct methanol (DMFC)	Methanol	Polymer	60–130°C	Portable electricity, transport?	?
Alkaline (AFC)	Hydrogen	Potassium hydroxide	80°C	Aerospace (transport)	1

Notes:
* Also mixtures of hydrogen and carbon dioxide. In combination with a 'reformer' these are also suited to hydrocarbons.

† Also called proton exchange membrane (PEMFC) or polymer electrolyte (PEFC).

Sources: Koppert et al. (1988); Schaeffer (1998); van der Hoeven (2001); http://www.ecn.nl/bct/fuelcellinfo/principle.en.html.

Hydrogen is considered to be a clean and versatile energy carrier that can contribute to the reduction of CO_2 emissions (if produced from renewable energy sources or nuclear energy). Fuel cells play a major if not exclusive role in the hydrogen economy (see Figure 5.1). Hydrogen can also be applied in internal combustion engines and turbines.

As early as the 1970s there were publications in the Netherlands about the potential role of hydrogen as the energy carrier of the future (see, e.g.,

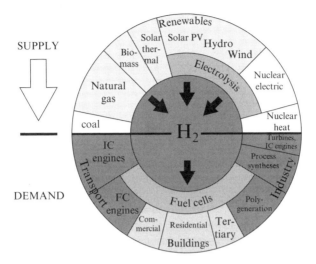

Note: Sizes of 'sectors' have no connection with current or expected markets.

Source: European Commission (2003).

Figure 5.1   *Hydrogen: primary energy sources, energy converters and applications*

Sanders, 1972; TNO, 1975; Lysen, 1977). As for the technology required for hydrogen production, the general thinking was in the direction of electrolysis of water with the use of electricity from nuclear energy and (in the longer term) also from wind and solar energy.[11] The presence of an intricate natural gas network in the Netherlands is regarded as a favourable factor for the introduction of hydrogen.[12] In addition to using hydrogen in fuel cells, other applications have also been considered, such as in standard combustion engines in cars and gas turbines in aircraft. In recent years the prospect of hydrogen as an energy carrier for the future has gained significant interest (see, e.g., Hoffmann, 2001; Dunn, 2002; Rifkin, 2002). The USA and the EU are spending large R&D funds on hydrogen, and Iceland has expressed its intention to become the world's first 'hydrogen economy'.

Actors

Dutch *research institutes* play a major role internationally in the field of fuel cell technology. In the 1950s, Broers and Ketelaar obtained good results at the University of Amsterdam, but the scope of their research was limited.[13] The Organisation for Applied Technical Research (TNO) began in the 1960s to engage in fuel cell research but pulled out again around 1970 (Schaeffer, 1998; van der Hoeven, 2001).[14] Since 1986 work on fuel cells has taken place at the Netherlands Energy Research Centre (ECN). Initially this mainly involved the MCFC, but that research was stopped in the late 1990s. Nowadays the focus lies mostly on SOFC and SPFC (PEMFC) and on related materials (see Table 5.1). In the academic world, fuel cell research takes place especially at the Delft University of Technology.

The involvement of Dutch *industry* in the development and production of fuel cells has been rather limited in the past. Starting in the late 1950s, the Dutch State Mines (later DSM) conducted some fuel cell research (Schaeffer, 1998), but the electricity companies (including their research institutes, such as KEMA and their umbrella organization SEP), and the gas trading company Gasunie kept their distance (Sanders, 1972; van der Hoeven, 2001). In 1987, steel producer Hoogovens decided to participate in the National Research Programme on Fuel Cells (see below under 'Policy'), but within a year the company had withdrawn from the programme.[15] Together with a Belgian partner, DSM (by then a major chemicals company) formed the joint venture, Elenco, which was dedicated to the use of AFCs for defence, aerospace and transport applications. However, Elenco went bankrupt in 1995 (ECN, 1999). Recently, DSM has again become involved in the development of fuel cells, as the company had access to a polymer that turned out to be very suitable for application in PEMFCs. For similar reasons, Akzo Nobel also became interested in PEMFCs in 1998. In that same year, however, the corporate research activities of Akzo Nobel were sold. Several employees of Akzo Nobel then founded Nedstack (van der Hoeven, 2001). Nedstack recently announced that, together with Akzo Nobel, it is going to build the largest fuel cell plant in the world at Akzo Nobel's chlorine factory in the Botlek area, near Rotterdam (Voermans, 2004).

In other countries the business community has always played a major role. This has involved, in particular, companies in the field of energy technology (such as General Electric, Westinghouse, ABB and Siemens). In the USA the gas industry has also conducted a large research programme intended for the development of a small electricity unit that private consumers could install at home to generate their own electricity (ECN, 1999). The electricity companies in the USA had heavily promoted the 'all-electric society'; the fuel cell programme of the gas companies can thus be viewed as a counter-offensive in the direction of an 'all-gas home' (van der Hoeven, 2001).

In recent years car producers in particular (with Daimler-Benz/ DaimlerChrysler in the lead) have been quite active in the field of fuel cell technology. In addition, there are specialized fuel cell producers (especially in North America), the Canadian company Ballard being the largest (Price-waterhouseCoopers, 2003). More than 500 companies are actively engaged in fuel cell technology around the world (Comyns, 2004).

International institutions such as the International Energy Agency (IEA) play an important role in international co-operation in the field of fuel cell technology. The EU is also quite active in this field (see also below under 'Policy'). The European Space Agency (ESA) has conducted fuel cell pro-grammes in connection with the energy supply for spacecraft.

Applications and niche markets
Fuel cells have the potential for wide-ranging applications. The first large niche market was, as mentioned earlier, the aerospace industry. In industry, Koppert et al. (1988) initially saw a market in situations in which there was a supply of hydrogen and/or a demand for direct current. Another appli-cation for which they saw good prospects for at the time was co-generation. For the longer term (15–20 years) large-scale electricity production from coal was considered (probably via coal gasification). Applications in the transportation sector were also seen as a potentially important area of application, but only in the more distant future.

Today the transportation sector, rather than large-scale electricity produc-tion, is where fuel cells are likely to experience their first major breakthrough. A significant factor that accelerated the application of fuel cells in vehicles was the 'zero-emission vehicle' programme in California, which was started up in 1990. Under this programme, 10 per cent of vehicles to come on to the market would by 2003 be totally emission free. Only electric vehicles (with a battery or fuel cell) were able to meet this requirement. The ZEV programme was amended later as the target turned out to be overly ambitious.

Car manufacturers are currently investing billions of dollars in the development of fuel cells. In particular, 'third-generation' cells (SOFC and especially SPFC/PEMFC; see Table 5.1) are considered eminently suitable for transportation purposes. The production costs (in 1998 approximately $6000 per kW, versus $60 per kW for an engine with internal combustion) are, however, still far too high to be competitive (Schaeffer, 1998).[16]

The learning curve
The cost of fuel cells has dropped significantly since the 1960s, from well above $100 000 per kW to well below $10 000 (ECN, 1999). A spokesman for the Dutch fuel cell producer Nedstack has pointed out that the price of fuel cells dropped by 50 per cent each year over the past five years and now comes to some €2000 per kW of invested peak capacity. He has predicted that this annual 50 per cent drop will continue for another three years and

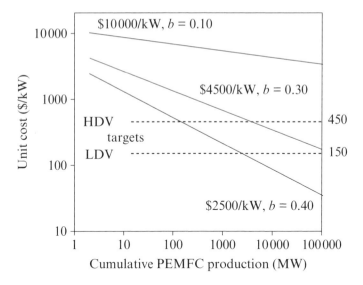

Notes: The curves are derived from the equation $C = aP^{-b}$, where C stands for the
production costs per unit, P for the cumulative production and a and b for constants. The
values of b are based on existing learning curves for gas turbines ($b = 0.10$) and PV cells
($b = 0.30$). The value $b = 0.40$ can be viewed as the upper limit. The values of a are based on
three estimates of current costs and an assumed current cumulative production of 2 MW.

Source: Rogner (1998).

Figure 5.2   Hypothetical learning curves for PEMFCs

will then level off (Voermans, 2004). This will mean that the price can drop
within three years to below €250 per kW.[17] A representative of Daimler
Benz expected in 1996 that the costs of a fuel cell motor would have come
down to $100−$150 by the year 2010 (a reduction by a factor of 300 with
respect to 1995) (Schaeffer, 1998, p. 405).

Research into the learning curve for fuel cells has focused mainly on
SPFC (PEMFC) for cars. In general, it involves analyses based on hypothetical learning rates, since the empirical data that are needed for the construction of learning curves for fuel cells are still scarce.

Rogner (1998) constructed hypothetical learning curves for PEMFCs
(see Figure 5.2). Based on these curves, he came to the conclusion that, even
under optimistic assumptions with regard to the parameters, a considerable
production volume would need to be reached in order to have the costs of
these fuel cells drop to a level at which they might compete on the car
market. In the case of heavy-duty vehicles (HDV in Figure 5.2) this would
be $450 per kW. This requires a cumulative production of between one

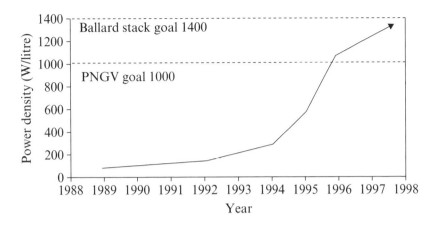

Note: PNGV = Partnership for Next Generation Vehicles, an R&D programme of the US government and the 'big three' of the American auto industry.

Source: Kalhammer et al. (1998).

Figure 5.3 *The historical development of the power density of Ballard fuel cells*

hundred and several thousand megawatts (by comparison, Rogner estimated the cumulative production through 1998 at less than 5 MW). For light-duty vehicles (LDV in Figure 5.2), the costs would need to drop to $150 per kW, which would call for a cumulative production of between ten thousand and several hundred thousand megawatts.[18] The estimated investments needed to reach such levels range from $180 million to $112 billion. These amounts relate to the additional investment costs that come on top of the costs of conventional vehicle propulsion.

Based on a learning curve analysis, Tsuchiya and Kobayashi (2002) expected the costs of PEMFC for cars to drop to a level comparable to that for internal combustion engines, provided that cumulative production reaches 5 million units. Schlecht (2003) came to the conclusion, on the basis of an assumed learning rate between 15 per cent and 25 per cent, and applying a game-theory analysis, that entering into alliances within the auto industry is the best way to introduce fuel cell vehicles on to the market. Van den Hoed (2004) illustrated the fast learning curve of fuel cell technology on the basis of the power density growth of Ballard fuel cells (see Figure 5.3).

Policy and institutional aspects
The Dutch government has contributed to R&D in the field of fuel cell technology for quite some time and, in doing so, has stimulated co-operation

between industry and institutes of knowledge. In 1984 a Fuel Cell Task Group was set up to advise on a National Research Programme. This programme came into effect in 1986, with a focus on the development of the MCFC. In 1990 the Netherlands signed the IEA Implementing Agreement on Advanced Fuel Cells. In the meantime the Dutch government has spent approximately €100 million on fuel cell technology. In an evaluation of this policy, van der Hoeven (2001) concluded that the ideal of an independent Dutch fuel cell industry had not been realized, but that the production of fuel cell components had by then appeared feasible for the Netherlands.

In the current Dutch Energy Research Strategy (EOS), PEMFC fuel cells belong to the *priority topics* of the energy R&D portfolio (EZ, 2003a), because they can contribute significantly to an economy based on sustainable energy. Furthermore, the Dutch knowledge base is strong (for further discussion, see Section 4.3.3). SOFC fuel cells belong to the 'import of knowledge' topics, which are options that can significantly contribute to a sustainable energy economy, but where the Dutch knowledge base is weak.

The EU also spends considerable research funds on fuel cell technology. In 2002 the European Commission set up a High Level Group on Hydrogen and Fuel Cells, which issued its final report in 2003 (European Commission, 2003). The European Hydrogen and Fuel Cell Technology Platform was then created, with the support of the EU.

5.2.2 Futures

For quite some time already, fuel cell technology has been viewed as a promising, clean and efficient option for electricity production in the future. They were already mentioned in a cursory way in the first 'Policy document on energy' (EZ, 1974). According to the 'Follow-up document on energy conservation', it was expected that by 2000 several megawatts of fuel cell power would be on hand (EZ, 1993). This was in line with statements such as those by Westinghouse that the commercial application of multi-megawatt systems would take place in the 1990s (Schaeffer, 1998). These forecasts did not come to fruition.

Expectations regarding the commercial application of fuel cells in vehicles were, according to Schaeffer (1998), also likely to be too high. As he puts it, it was not to be expected that a fuel cell car would appear in the showrooms in 2004, but ten years later that might well be the case (Schaeffer, 1998, p. 493). In a similar vein, van den Hoed (2004, p. 287) states that commercial fuel cell autos can be expected in 10–15 years at the earliest, considering the great technological uncertainties that surround them.

The report by the High Level Group on Hydrogen and Fuel Cells can be read as the EU's scenario with regard to fuel cells for the future (European

Commission, 2003). It plots the contours of a 'hydrogen-oriented economy' and shows the route that might lead us there. Up to 2020 the emphasis will be on research, development, test projects and niche markets, while after 2020 the major breakthrough of fuel cells in mobile and stationary applications could take place.

The Dutch government also counts on a breakthrough of fuel cells in the long term. According to its report on 'Energy and Society in 2050', the Ministry of Economic Affairs sees fuel cells playing an important role in transportation in three of the four scenarios described (EZ, 2000b).

Nonetheless, it is far from certain that fuel cells will actually play a large role in energy supply. The uncertainty about the future role of fuel cells is illustrated by statements by Robert Lifton, director of the American fuel cell producer Medis Technologies, at the World Economic Forum in Davos (January 2004). He said that public authorities would do better to focus on energy conservation and efficiency, since fuel cells are difficult to upgrade to the power that is needed for passenger cars or computers. In response to this, ECN stated that Lifton sketched a far too negative picture about the development of fuel cells, especially for the transportation market. According to ECN, the fact that commercial products with fuel cells are not yet available has to do with series production and infrastructure.[19] In other words – contrary to petrol engines, for example – fuel cells do not or cannot yet profit from economies of scale, network externalities, complementary technologies and learning curve effects.

What type or types of fuel cells will come on to future markets cannot yet be predicted. In view of their diverse properties, it is not unlikely that different types will coexist for some time, depending on the field of application.

5.2.3 Driving Forces and Barriers

This section discusses the principal factors that impact the development and application of fuel cell technology. These factors correspond with the aspects listed in Table 2.1.

Diversity
The research into and development of fuel cells is marked by great diversity. Large as well as small companies from various *business sectors* are involved. There is also great diversity in *techniques* and *products*. In addition to different types of fuel cells and various application possibilities, there is also a variety of energy sources that can be used. Hydrogen can be produced in many different ways: from fossil fuels but also with the help of renewable energy sources and nuclear energy.

Diversity in R&D *strategies* applies especially to the different business sectors and applications. Within the auto industry, on the other hand, there is much homogeneity in R&D strategies (see van den Hoed, 2004, chapter 6).

Innovation
In some fields of application, *combinations* of fuel cells with 'traditional' forms of energy conversion appear to hold the greatest promise. In electricity production, for example, the expectations are high for SOFC fuel cells that have a small gas turbine added to them (Schaeffer, 1998). Furthermore, in the absence of a hydrogen infrastructure, there is research into the possibilities of combining fuel cells with traditional (fossil) fuels. This may be feasible by using reformers, which convert fossil fuel into hydrogen (e.g. aboard a vehicle).

Co-operation between different types of companies and research institutes plays an important role in the development of fuel cell technology. This applies to combinations of expertise in the field of fuel cell technology and applications (e.g. Ballard and DaimlerChrysler), but also in combinations of small companies that possess knowledge and ideas with large companies that have funding (e.g. Nedstack and Akzo Nobel). We also see that with the rise of the PEMFC, companies with knowledge of polymer technology (such as Akzo Nobel and DSM) are getting involved in fuel cell R&D.

Niche markets for innovative fuel cell applications include the aerospace industry and the market for zero-emission vehicles (enforced by regulations in California; see below). Within the vehicle market there are also other potential niche markets such as vehicle fleets (e.g. buses and taxis). As long as there is no widespread network of hydrogen gas stations, driving on hydrogen is mainly of interest for fleet vehicles of this kind which, after a certain distance, always return to the same location, where they can take hydrogen on board.

As we saw in Section 5.2.2, there is no uniform and widely accepted *vision* of the future role of fuel cells in energy supply. Attempts have been made, however, to arrive at such a vision; for example, in the report of the High Level Group of the European Commission.

The selection environment
The *technical possibilities* of fuel cells are considerable, but the *economic* feasibility of large-scale application is yet to be proven.

Geographical characteristics play a role, especially in Iceland, where the abundance of cheap possibilities for the production of electricity from renewable sources (terrestrial heat and hydropower) represents a favourable condition for the introduction of hydrogen and fuel cells.

In most *markets*, fuel cells are not yet a competitive option. Opinions differ on the answer to the question about which market will be the first to be ripe for fuel cells. Although most of the attention is directed at the auto market, it is also quite possible that a market such as portable electronics will win the race.[20] Sanders (1972) noted that fuel cells for a specific group of companies can already become attractive in an early phase; namely, for companies that require a low-voltage direct current for, for example, electrolysis and galvanization. Expensive battery chargers would then no longer be needed. These companies could thus become a niche market for fuel cells. Koppert et al. (1988) also alerted us to a potential market in situations where there is a supply of hydrogen, such as in chemical processes that produce hydrogen as a by-product or waste. An example of this is chlorine production (see Voermans, 2004).

Rogner (1998) identifies three factors that are critical to lifting a technology from the R&D phase to the commercialization phase; namely, niche markets, government policy and early adopters. Of these three, *government policy* (especially on environmental issues) is, according to him, probably the most important in the case of hydrogen technology. Van den Hoed (2004) also views governmental regulations (such as the Californian requirements regarding zero-emission vehicles) as a significant factor in the increasing interest shown by the auto industry in the fuel cell.

In the Netherlands and the rest of Europe, government policy can generally be viewed as conducive to fuel cells. The liberalization of the electricity market may well provide options for new distributed electricity providers. This could have a favourable impact on the development of fuel cells. On the other hand, persistent lock-in of the current electricity system is also conceivable. Also there are competing technologies that can profit from the liberalized climate (such as distributed co-generation). The relatively clean and efficient alternatives obviously face the handicap that they must compete with traditional forms of electricity production, where the external costs are not fully reflected in the price.

Bounded rationality
The limited *time horizon* that companies apply in their investment decisions has clearly impacted fuel cell R&D in the Netherlands. Hoogovens, for example, stepped out of the National Research Programme because of the excessive payback period and the high level of uncertainty (see note 15).

In some respects, the introduction of fuel cells requires a break with current *routines*. While we are now used to simply inserting a plug into a socket, we may in future be able to produce electricity directly using a fuel cell. This leads to a variety of questions: on the type of fuel cell needed and the required capacity; whether the same fuel cell can be used for different purposes; whether an electricity surplus can be sold; and so on. Current experience with the introduction of co-generation in industry and small-scale energy production by private persons (wind turbines, solar panels) is quite useful in this regard.

Imitation behaviour can also be noted. Following the lead taken by DaimlerChrysler, many car producers have started to develop fuel cell cars, so as not to miss the boat (see Schaeffer, 1998, p. 402ff.; van den Hoed, 2004, chapter 6).

Path dependence and lock-in

Increasing returns to scale will occur in fuel cells, mostly in the form of learning effects and economies in production. In the application of fuel cells there are only limited economies of scale, since fuel cell systems are constructed in a modular fashion ('stacks').

Especially in the field of transport applications, fuel cells must face the extensive *lock-in* of fossil fuels and the combustion engine. Lock-in situations probably also exist in other areas, such as in batteries (in various off-grid applications, where fuel cells might be an alternative).

Co-evolution

The development of fuel cells is closely aligned with developments in chemistry, especially in the fields of polymers and catalysis. Large-scale introduction of fuel cells will have a major impact on the total energy system. Fuel cells fit well into a distributed energy system, where fuels are converted into electricity and heat at the very location and point in time where they are required. However, this requires a fuel infrastructure, which leaves the question of which fuels will dominate this infrastructure. Candidate fuels include hydrogen, natural gas, methanol, petrol and maybe others as well.

5.2.4   Conclusions

Fuel cells already have a long history. They have rekindled interest in view of their potential benefits to the environment and their energy efficiency. However, there are still great uncertainties about the economic viability of this technology for the long term. In addition, there are still many technical problems to be overcome.

Both Dutch and EU policy on fuel cells has until now focused heavily on stimulation of R&D. While in the case of the Dutch Research Programme this has not led to the hoped-for results, it has clearly contributed to the position that the Netherlands has acquired in the field of fuel cell technology.

From the viewpoint of evolutionary economics, the time may have come to devote more effort to the application and spread of the technology. The course that has yet to be followed along the learning curve before any commercial and large-scale applications become possible is still long. Following this line of reasoning, it is also important to stimulate niche and other markets so that the pace of cumulative production of and experience with fuel cells can pick up. The relevance of niche markets is evidenced by the major role that the Californian zero-emission vehicle requirements have played in the rise of the fuel cell in the car industry. The potential role of the public sector as launching customer could also be looked into.

With a view to the sustainability of energy supply, the question arises of whether the application of hybrid systems (such as combining the fuel cell and the combustion engine and the possibility of using fossil fuels) will not lead to a lock-in of fossil technology. For that reason, Dunn (2002) pleads for government initiatives to create a hydrogen infrastructure, so that the big step to a hydrogen economy can be taken without remaining with one leg stuck in fossil fuel technology. Evolutionary economics seems unable to provide an unambiguous solution to this problem. On the one hand, lock-in resulting from network effects and complementarity plays a large role (which may be reason enough to plead for public support for an alternative network). On the other, hybrid systems can often play a useful role as an intermediate phase in the development of a new technology.

In summary, it may be said that from an evolutionary-economic viewpoint, the large diversity (in companies, techniques and product features), the interaction and collaboration between different players and public policy (zero-emission regulations) have been (and continue to be) significant factors in the development of fuel cell technology. However, to allow the fuel cell to benefit from economies of scale and to cut through the lock-in of current technology (in particular, the combustion engine), stronger stimuli will be required from the selection environment. This may call for new environmental regulations and stimulation of niche markets.

5.3   NUCLEAR FUSION

5.3.1   History and the Current Situation

Introduction
Research into nuclear fusion started around 1920, when British physicists discovered that during the fusion of hydrogen atoms into helium, mass is converted into energy. Nuclear fusion as a potential energy source gained

interest after the first hydrogen bomb tests in the 1950s. The idea was that if the enormous amount of energy generated during nuclear fusion[21] could be released in a controlled way, the world would then possess an almost unlimited source of energy. In addition, this method of energy production would not involve the dangers of nuclear explosions or the problems of radioactive waste. In practice, however, keeping these promises has turned out to be more difficult than originally thought.

In simple terms, the problem of controlled nuclear fusion with a net energy yield is that a combination of extremely high temperature (millions of degrees Kelvin, in which hydrogen enters a plasma state) and extremely high density must be maintained for a sufficient period of time. This requires an installation (called a *tokamak* or *torus*) in which very strong magnetic fields must restrain the plasma. Such installations have been constructed in several countries.[22]

At the European level, a start was made in 1979 with the construction of the Joint European Torus (JET) at Culham in the UK. This went into operation in 1983. Over the years significant progress was recorded here as well as in other tokamaks. The *triple product* is a measure of the performance of fusion plasma, expressed as density × temperature × time. The triple product that can be realized experimentally has increased by a factor of 100 000 in 30 years (see Figure 5.4). In 1991, 1 MW of fusion power was realized for the first time, and in 1997 a power level of 4 MW was maintained for four seconds (ITER, 2004).

The EU is now working together with other countries (China, Russia, Japan, India, the USA and South Korea) on a new project called the International Thermonuclear Experimental Reactor (ITER). This is intended to lead to the realization of a tokamak that is significantly larger than the tokamaks built to date. If everything goes as intended, this experimental nuclear fusion reactor will produce several hundred megawatts, a quantity of energy that is comparable to the output of a regular power plant. Recently, it has been decided that ITER will be located in Cadarache in France. The investment costs of ITER are estimated at €5 billion.

The actors

The European Community for Atomic Energy (Euratom) is the driving and coordinating force behind nuclear fusion research in the EU. Since Euratom's founding in 1957, nuclear fusion research has always been part of the European research programmes. Table 5.2 presents an overview of the amounts that the EU has spent on nuclear fusion research since 1990. During the period from 1974 to 1998, the countries belonging to the International Energy Agency (IEA) spent a total of $26.8 billion on R&D in the field of nuclear fusion.[23]

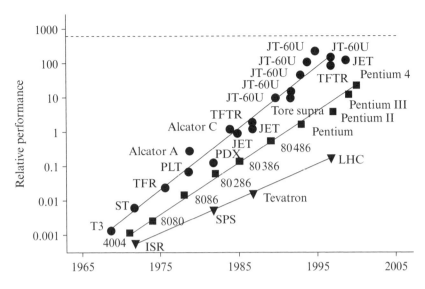

Notes: Since the early Russian T3 tokamak, the performance of fusion plasmas has doubled every 1.8 years (●). The performance of fusion plasmas is defined in terms of the triple product (density × temperature × time). This triple product compares favourably with the doubling of the energy of particle accelerators every 3 years (▼), and the doubling of the number of transistors on a chip every 2 years (■). The dashed line at the top shows the performance expected with ITER.

Source: Hoang and Jacquinot (2004).

Figure 5.4   How fusion experiments have kept pace with other high-tech developments over the past 30 years

Table 5.2   Expenditures on nuclear fusion research under the EU framework programmes since 1990

Programme	Budget
Third Framework Programme (1990–4)	ECU 458 mln
Fourth Framework Programme (1994–8)	ECU 794 mln
Fifth Framework Programme (1998–2002)	EUR 788 mln
Sixth Framework Programme (2002–6)	EUR 750 mln
Total 1990–2006 (1 ECU = 1 euro)	EUR 2790 mln

Sources: EC Bulletin 9/1990; Lyons (1996);
http://europa.eu.int/comm/research/fusion/fusion3.html (20 February 2004);
http://europa.eu.int/comm/research/energy/fu/fu_rd_en.html (20 February 2004).

Scientific institutes are the prime actors in the field of nuclear fusion. In the Netherlands, these include especially the FOM Institute for Plasma Physics 'Rijnhuizen', founded in 1958, which contributes to the European nuclear fusion research programme.[24] Researchers at Rijnhuizen work on such projects as JET and, along with German and Belgian colleagues, on the TEXTOR fusion experiment in the Kernforschungszentrum Jülich. Rijnhuizen also collaborates with a number of Dutch universities in the Centre for Plasma Physics and Radiation Technology (CPS).

In addition to particle physics, many other disciplines are involved in the design and construction of nuclear fusion reactors. To realize the required electromagnetic fields, for example, calls for knowledge and technology in the field of superconductivity. The walls of the reactor require advanced materials, for which research is taking place at the Nuclear Research and Consultancy Group (NRG) in Petten.

The role of the *business community* in the development of nuclear fusion as a source of energy has until now been confined to that of actual or potential supplier of materials and installations. A company such as Holec has, in the past, received contracts under the European fusion programme and is also authorized to deliver components for ITER (van Hilten et al., 1996). The *energy companies* have never shown much interest in nuclear fusion. In their report on the 'future energy situation in the Netherlands' (VDEN, 1980), nuclear fusion is labelled as being futuristic.

Applications and niche markets

At the moment there are no commercial applications of nuclear fusion, and this situation may prevail for quite some time. Considering the large economies of scale, the high investments and the relatively low fuel costs, it is already clear, however, that nuclear fusion will most probably only lend itself to large-scale applications such as base load capacity in electricity supply.

Policy and institutional aspects

Government policy with regard to nuclear fusion (in the Netherlands, the EU and elsewhere) has until now focused almost exclusively on stimulation of research and development. The Dutch Ministry of Economic Affairs does not provide any direct subsidies to nuclear fusion research. The government's position is that in view of its international significance, nuclear fusion research must be financed within the context of the EU.[25] There are no indications of much consideration being given in the policy arena to the implications of a possible breakthrough of nuclear fusion for the overall energy system.

5.3.2 Futures

For more than 30 years, many people have viewed nuclear fusion as a potentially clean, safe and practically inexhaustible source of energy. From the very start it was clear, however, that nuclear fusion would not be an energy source for the near term.[26] According to the Dutch 'Energy policy document' of 1974, 'at the earliest a contribution to the supply of energy from nuclear fusion can be expected at the start of the next century' (EZ, 1974, p. 158). The Dutch nuclear fusion expert Braams (1982) noted that it was difficult to imagine a development that would lead to the commercial application of nuclear fusion for the production of electricity before the year 2020. By now developments are such that the commercial application of nuclear fusion is not expected before 2050. In addition, it is still far from certain that the application of nuclear fusion for the supply of energy will be technically possible and economically feasible. Thus, in most long-term energy scenarios and visions, nuclear fusion hardly plays a role.

An estimate given by Lako (2001) of the investment costs of a double 1500 MW nuclear fusion power plant in the year 2100 amounts to €3000 per kW. A commercial fusion plant of 1000 MW could produce electricity at a cost of nearly 7 eurocents per kWh. That would not allow the technology to compete with alternative base load options such as coal-fired power plants, unless a stringent policy is applied to reduce CO_2 emissions. If that happens, nuclear fusion could then account for a significant portion of the electricity supply in 2100, especially in scenarios where the demand for energy is growing sharply, or where the availability of oil and gas is limited, and/or nuclear fission is reduced due to lack of public acceptance.

Tokimatsu et al. (2003) also come to the conclusion that nuclear fusion can only be competitive if stringent CO_2 requirements apply (below 550 ppmv). The break-even price (in US currency) would then lie between 6.5 and 13.5 cents per kWh in 2050 and between 9 and 30 cents in 2080. According to the authors, nuclear fusion reactors that provide electricity at a cost of 7–13 cents per kWh may become available from 2050, and therefore they see good prospects for their introduction during the 2050s. By 2100 nuclear fusion could then account for a maximum of 30 per cent of total electricity production.

Any commercial nuclear fusion reactors that might come into operation during this century will probably be of the tokamak type. Meanwhile, work is being conducted on other types of plasma containment devices (see note 22), but these technologies lie a full generation behind the tokamak.

5.3.3   Driving Forces and Barriers

This section discusses the principal factors that impact the development
and application of nuclear fusion technology and corresponds with the
aspects listed in Table 2.1.

Diversity
There is little diversity within nuclear fusion technology. As a technique for
plasma containment, the tokamak is currently the furthest developed.
Other technologies are still very much in an early phase. In view of the enormous investments that nuclear fusion calls for, the technological diversity
will most likely always remain limited.

Innovation
Nuclear fusion research largely takes place within an exclusive circle of
insiders spread around the world. The mutual *exchange* of knowledge and
experience is intensive. To illustrate this, even during the Cold War, the
Russians, Europeans and Americans kept each other informed about their
results (the tokamak, for example, is a Russian invention).
    Venture capital or *niche markets* do not yet figure at this stage. The
development of the technology depends entirely on government funding.
    As mentioned in Section 5.3.2, an unambiguous *vision* of the role of
nuclear fusion in the supply of energy does not exist. The only consensus
relates to the expectation that nuclear fusion will not play any role in commercial energy supply before 2050.

The selection environment
For the time being, the *technical* viability of nuclear fusion as a source of
energy will first have to be demonstrated. The next question is whether this
technology can also be made feasible in an *economic* sense. The answer to
this depends on the extent to which stringent CO_2 *policy* will apply
(meaning that traditional forms of power production with fossil fuels will
have to be abandoned), and also on the development of competing technologies that do not involve CO_2.

Bounded rationality
The investment of large sums of public money in nuclear fusion
research could be justified by referring to the limited *time horizon* of private
investors. Sinking billions of euros into a technology that might (or might
not) become profitable in 50–100 years is not something that private companies are anxious to do. On the other hand, for public authorities, support
for nuclear fusion research, and the prospect of an inexhaustible and clean

energy source, is apparently sufficient to cut through the short-term thinking. A final consideration is that nuclear fusion is a technology that differs altogether from prevailing energy technology, so that the possibilities to build on existing *routines* are few.

Path dependence and lock-in

The investments in nuclear fusion technology in the past, which are seen as being irreversible, will determine the future development to a great extent. This means strong path dependence. For example, the choice of a specific plasma containment technique (the tokamak) will determine the basic design of nuclear fusion reactors for many years to come.

Economies of scale in the production process play a major role in nuclear fusion. These have a mainly technical background. The *learning effects* are also significant: the more fusion reactors there are, the more experience can be exchanged (but the number of reactors will always remain limited in view of the minimum required scale size).

Like all 'alternative' forms of energy supply, nuclear fusion will have to cut through the current *lock-in* in fossil fuels and nuclear fission in order to be successful. However, nuclear fusion can be fitted well into the existing systems of large-scale, centralized electricity production. With economies of scale, a lock-in in nuclear fusion technology is conceivable in the long term, even though this will only relate to the production of base load electricity.

Co-evolution

The development of nuclear fusion takes place outside the mainstream of energy technology, which means that there is only limited interaction with other energy technologies. With disciplines such as physics and materials science, however, there are strong ties. Furthermore, fusion research is expected to lead to spin-off effects in the fields of cryogenic technologies, microwave technology, superconductive materials, remote handling and the industrial use of plasmas.[27]

If nuclear fusion turns out in time to be the almost inexhaustible and clean energy source that has often been promised, this will undoubtedly have great repercussions for the nature and scope of society's energy use. Any pronouncements in this regard are for the moment strictly speculative.

5.3.4 Conclusions

For decades, nuclear fusion has been seen as an energy source that holds great promise for the long term. Heavy government funding has resulted in a good deal of progress in the development of this technology, but the road to commercial application is still long. The uncertainty about whether nuclear fusion will ever provide energy at a competitive cost is great.

From an evolutionary-economic perspective, the substantial financial support for nuclear fusion research may be justified in the light of the limited time horizon of private investors. It should be noted, however, that this could happen at the expense of providing financial possibilities for the development of other energy technologies, which, for the same reason, would not get off the ground. The diversity of technologies is thus likely to be limited. The development of nuclear fusion is isolated from other energy technologies. This means that there is little room for cross-fertilization and co-operation, as a result of which the effectiveness of R&D stimulation (in terms of spillover effects) is likely to be lower compared to other, less isolated, fields.

The application of nuclear fusion as a practical energy source is still very remote. For now this means that there is no possibility of stimulating the technology through the creation of niche markets or a favourable selection environment. Lastly, it should be noted that nuclear fusion cannot be applied on a small scale, so that this technology is not compatible with distributed systems of electricity production. This also has a limiting effect on the diversity of energy technologies.

All in all, nuclear fusion is a problematic energy technology from an evolutionary-economic viewpoint, especially because of the lack of diversity in various respects. Whether the selection environment will be conducive for this technology in, say, 50 years time (when nuclear fusion is likely to be ripe for the market) is impossible to say. If not, then we can always comfort ourselves with the thought that wastage is an inevitable phenomenon from the viewpoint of evolutionary economics (see Section 2.4.2).

5.4 PHOTOVOLTAIC ENERGY (PV)

5.4.1 History and the Current Situation

Introduction

Photovoltaic cells (PV cells) convert sunlight directly into electricity. Their functioning is based on the principle that the photons in light are able to release electrons in semiconductors (such as silicon), thereby creating an electrical voltage.

PV technology has its roots both in quantum mechanics and in the semiconductor revolution of the 1950s. The first efficient solar cells were produced in 1954, and the first commercial application (in aerospace) dates from 1958. The oil crisis of the 1970s gave a new stimulus to PV research. Originally this technology was used mainly in locations where no electricity

Table 5.3   An overview of PV technologies

Type	Subtype/material	Characteristics	Actual yield	Applications
Silicon wafers	Monocrystalline Si Multicrystalline Si	Oldest generation; highly developed	> 20%	Wide
Thin film (second generation)	Amorphous Si (with Ge)	Production at low temperatures	6–7%	Among others in calculators
	Polycrystalline semiconductor compounds (CdTe, CuInSe$_2$)	The required materials can be a bottleneck	12–19%	Various, including modules (still in pilot phase)
	Polycrystalline Si cells	Relatively high capacity to absorb light	> 10%	In development
	Nanocrystalline organic pigments*	Process roughly similar to photosynthesis	Unknown	In development

Note: * This type is also referred to as the third generation.

Source: Green (2000).

grid was available, but the past few years have seen a significant growth in applications that are grid-connected (stimulated by subsidized programmes). By the end of the 20th century, PV industry revenues had passed the $1 billion mark (Perlin, 1999; Green, 2000). The cumulative installed PV capacity has meanwhile increased to well above 1000 MW$_p$.[28]

Table 5.3 presents an overview of both current and developing PV technologies. The first generation, which is still dominant, is based on the use of silicon wafers. The second generation consists of thin-film technology, which has the benefits of lower material costs and greater suitability for mass production. Within this technology there are several subcategories that are already, or will soon be, commercialized. It is expected that the next ten years will make clear which of these subcategories (all of which have pros and cons) will become dominant. Meanwhile the contours of a third generation are becoming visible. Here the efficiency with which sunlight is converted into electricity (currently subject to a theoretical maximum of 33 per cent) can be raised to approach the thermodynamic maximum of 93 per cent. This is made possible by combining different types of cells, each of which is optimized for a different part of the spectrum (Green, 2000). Work is also being conducted on solar cells with a yield that is not as high but which involve lower production costs; for example, by taking synthetic materials as a base (de Wit, 2004).

The actors
The technology used for the production of most solar cells leans heavily on the *micro-electronics industry*. This allows the PV sector to benefit from the economies of scale that are achieved in this much larger industrial sector. Furthermore, less stringent quality criteria apply for silicon crystals used in solar cells, so that raw materials and silicon wafers that do not meet the specifications for micro-electronics can be used. The largest players in the PV market include the Japanese electronics companies Sharp and Kyocera. As the PV industry becomes more mature, it will inevitably make increasing use of technology that has been optimized for its specific requirements, such as multicrystalline silicon wafers (Green, 2000).

Several *oil companies* had already become engaged in the PV field in an early phase of development.[29] Some of them, such as the former Mobil, later pulled out of the PV market, but Shell and BP are now among the major players. *Other energy companies*, such as the German company RWE, also started to produce PV systems. In addition, energy companies are obviously involved in the PV market as customers or by initiating and stimulating grid-connected PV systems. Some energy companies (Nuon in particular) present themselves as PV supporters through their sponsorship of projects such as the PV-propelled vehicle race across Australia.

Specialized PV suppliers are, in general, rather small, except for the Spanish company Isofotón and the American company Astro Solar. Most of these companies have their origins in the academic world (universities of technology).

In the Netherlands, Shell Solar was until recently the only producer of PV cells and modules, but it was joined in 2002 by Philips Solar Energy and Logic Electronics. Akzo Nobel has plans to start with the production of PV systems (IEA, 2003). Shell Solar, on the other hand, decided in 2002 to close its production plant in Helmond as part of a reorganization owing to excess capacity on the PV market.[30]

The most advanced PV technologies are applied in the *aerospace industry*. Dutch Space, which originated out of Fokker, produces solar panels for aerospace purposes and is the market leader in this field in Europe.

In addition to the producers of PV cells and modules there are also specialized providers of other PV system components, such as inverters, batteries, accumulators and technical constructions. Consultants, and engineering firms and contractors serve as the interface between the producers and potential users of PV systems and for the conversion of technological know-how into concrete applications. These users include companies and institutions in many different sectors, plus private individuals. As the number of PV applications integrated into buildings grows, so will the role of architects and construction companies, especially when PV elements are used as building components. Banks and other financial institutions lend money to finance PV systems.[31]

Governments play a key role through their financial and other support to research and development, pilot projects and investments, and tax incentives for the production and use of PV-based electricity (see also below under 'Policy and institutional aspects'). They are also in a position to ensure a regulatory framework and an institutional setting that is favourable for PV. In the Netherlands this applies in particular to the Ministry of Economic Affairs (responsible for energy policy, including sustainable energy), with Senternovem as the executive body to administer various subsidy schemes. Local public authorities are also often involved in the initiation or support of PV projects.

Universities and other research institutes play a major role in the PV arena. In the Netherlands, ECN has been engaged in research and development on PV since 1990. At ECN Solar Energy, over 50 specialists are working to develop solar cells and panels with a higher conversion yield and a better price/performance ratio. At another branch of ECN, DEGO (the Dutch acronym for 'sustainable energy in built-up areas'), work is taking place on the electrical and constructional integration of PV systems in buildings and on the combination of thermal and electric solar energy systems (ECN, 2004). In the Dutch Polymer Institute (DPI, one of the four top technological institutes in the Netherlands), some 20 trainee research assistants and postgraduates are working full time on polymer solar cells (de Wit, 2004).

International organizations are very important interfaces in the co-operation and exchange of expertise between countries. These include the International Energy Agency, with its Photovoltaic Power Systems Programme, and the EU (which is highly focused on renewable energy through the Sixth Framework Programme for Research).

Applications and niche markets
In addition to the aerospace industry, there are four major application fields for PV (IEA, 2003):

- *Off-grid domestic*: Small, isolated rural communities, especially in developing countries, represent an important niche market. Off-grid

PV systems form an economically feasible alternative to enlargement of the electricity grid where the distance to the existing grid exceeds one or two kilometres.

- *Off-grid non-domestic*: In addition to aerospace (the oldest niche market for PV), there is a great variety of applications where small electricity volumes represent high value, such as telecommunications, water pumps, vaccine cooling, navigation tools, airline signal lights and meteorological recording equipment.
- *Grid-connected distributed*: The increase in PV applications in buildings that are connected to a grid is a relatively new phenomenon. This involves small systems (0.4–100 kW) that handle the supply of electricity to the building itself, with any excess flowing back to the grid. These systems have several advantages: reduction of distribution losses, no extra capacity requirements, the possibility of integrating PV panels into the roofing of houses, and generally no need for energy storage.
- *Grid-connected centralized*: These systems are installed as an alternative to conventional centralized electricity production or as reinforcement of the distribution system. Demonstration projects that currently operate in various countries provide experience for the construction, operation and performance of such systems.

Off-grid systems are thus economically feasible in many cases even now. Grid-connected systems usually depend on government subsidies or on initiatives by companies and private persons who are not primarily driven by financial or economic considerations.

By the end of 2002, cumulative PV power amounting to 1330 MW had been installed in the 20 countries that belong to the Photovoltaic Power Systems Programme of the IEA (IEA-PVPS), with nearly half of this power in Japan. These 20 countries account for more than 90 per cent of worldwide PV production. Since 1994, installed capacity has grown at an annual rate of 20–40 per cent (IEA, 2003).

In the Netherlands too, installed PV power has grown substantially in the past ten years; at the end of 2002 it amounted to 26 MW$_p$ (see Figure 5.5). This growth related especially to grid-connected distributed systems. In installed PV power per capita, the Netherlands is now fifth worldwide (with 1.6 W$_p$ per capita), after Japan (5 W$_p$ per capita), Germany, Switzerland and Australia. The largest PV roof-situated project in the world, which accounted for 2.3 MW, was realized in 2002 at the 'Floriade' horticultural exhibition (Lysen, 2003). The Netherlands also has the only 'city' in the world where the entire electricity demand is met by PV; namely, the miniature city of Madurodam (van Beek et al., 2003).

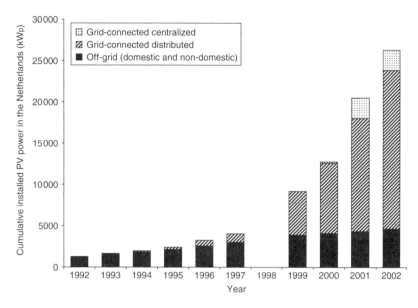

Source: van Beek et al. (2003).

Figure 5.5 *The cumulative installed PV power in the Netherlands, by category*

The learning curve

Harmon (2000) constructed a learning curve for PV modules for the 1968–98 period, during which the worldwide cumulative installed power of PV modules grew from 95 kW to 950 MW. The costs during this same period dropped from $90 to $3.50 per W_p.

The learning curve can be expressed in a general sense as:

$$Cost(CUM) = Cost_o * CUM^b$$

in which:

Cost (CUM)	= the cost per unit as a function of cumulative shipments
$Cost_o$	= the cost of the first unit shipped
CUM	= cumulative shipments over time
b	= the experience index

The learning rate (*LR*) is defined as follows:

$$LR = 1 - 2^b$$

The learning rate refers to the percentage by which the costs go down when cumulative shipments are doubled.[32] In the case of PV modules this amounts

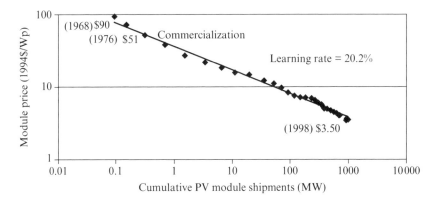

Source: Harmon (2000).

Figure 5.6   The learning curve for PV modules, 1968–98

to 20.2 per cent. The related learning curve is presented in Figure 5.6, in a double-logarithmic diagram.

Poponi (2003) arrived at a learning curve with a learning rate of 25 per cent for 1976–2002 (partially on the basis of other data). This high rate is strongly impacted, however, by the very high prices at the start of this period, when only few PV systems were sold. For the period 1989–2002, Poponi arrived at a learning rate of approximately 19.5 per cent, very close to that of Harmon.

The PHOTEX project, funded by the Fifth Framework Programme of the EU, focuses on the further development of PV learning curves.[33]

Policy and institutional aspects
Because of the lack of sunshine in the Netherlands and the availability of a detailed electricity grid, the opinion prevailed for a long time that PV was not a realistic option for the country. PV was mostly seen as an option for off-grid applications. In the first National Research Programme for Solar Energy (NOZ), conducted between 1978 and 1982, PV was therefore hardly mentioned (Knoppers and Verbong, 2001). During the second phase of the NOZ (1982–5), PV got a bit more attention, and the third phase (1986–90) included a separate PV programme of NLG 12 million (€5.5 million). During the 1990s the Dutch government spent some NLG 50 million (€23 million) per year on solar PV (EZ, 1999). Most of this money went to research and development and to demonstration projects.

Innovative PV projects have been until recently entitled to subsidy under the Decree on Energy Programme Subsidies – Sustainable Energy in the Netherlands (BSE–DEN), with SenterNovem as the executive agency. The subsidy instruments are being adjusted within the context of the Energy Research Strategy (EOS) (EZ, 2003a). EOS has identified multicrystalline and thin film solar cells as the spearheads of the energy R&D portfolio, as these are options that can contribute significantly to an economy based on sustainable energy and as the Netherlands has a strong position in terms of related expertise. Investments in PV are stimulated by means of such incentives as the Energy Premium Regulation, the Energy Investment Tax Credit, the Vamil (arbitrary depreciation of environmental investments), the CO_2 reduction plan and the Green Investment Regulations.

Until 2003 the use of green electricity (in particular, from wind, PV and biomass) was stimulated through a zero energy tax rate. This regulation is now replaced by a producer subsidy based on the MEP regulation (Environmental Quality of Electricity Production). Specific stimulation of PV through high feed-in rates, as in Germany, does not occur in the Netherlands.

In recent years the Dutch government has placed more emphasis on the cost-effectiveness of energy subsidies, expressed, for example, in kilograms of CO_2 reduction per subsidy euro. This is to the detriment of PV, since other technologies lead to lower CO_2 emissions at lower cost, at least to date.

In 1997 a PV agreement was signed by the Ministry of Economic Affairs, energy producers and other market players. The purpose of this agreement was to reach 7.7 MW of PV capacity by the year 2000. This target was fully achieved.

PV is also stimulated at the international level. In its framework programmes, the EU spends large sums on R&D in the field of renewable energy and has special funds for practical experiments and model projects (such as the Altener programme). In the past ten years the European Union has spent more than €200 million on R&D projects in the PV field. Recently the European Commission established a Photovoltaics Technology Research Advisory Council (PV-TRAC). This council published its long-term vision on PV technology in September 2004 (European Commission, 2004a). The countries belonging to the IEA spent a total of $5.2 billion on R&D in the PV field between 1974 and 1998.[34]

An important institutional condition for the development of the PV market is the existence of standards and codes. With this in mind, the International Electrotechnical Commission (IEC) set up a technical committee in 1981, which by the year 2003 had issued 26 international standards. At the European level CENELEC is responsible for setting PV standards and codes (IEA, 2003).

5.4.2   Futures

In most scenarios and visions, the contribution of PV to the worldwide energy supply for the short and medium term is not estimated to be very high. Criqui and Kouvaritakis (2000), for example, expect 15 TWh of electricity generation to be from solar energy by the year 2030, which accounts for a mere 0.04 per cent of global production. Likewise, the IEA (2002) does not expect any major contribution to electricity generation from renewable sources in the next few decades, despite strong growth. It mentions 7 per cent in 2030 for all non-hydro renewables combined, and it may be assumed that PV represents only a minor portion of this. Bauquis (2003) arrives at only 3 per cent in 2050 for the same parameter. Only in the two scenarios of Shell (2001) does the share of non-hydro renewables in the energy supply in 2050 substantially exceed 10 per cent.

According to the PV-TRAC report (see Section 5.4.1), PV could account for 4 per cent of electricity generation in 2030 (European Commission, 2004a). PV-TRAC pleads for the establishment of a European PV Technology Platform to stimulate and coordinate PV-related initiatives, programmes and policy.

As for the Netherlands, the expectations for PV have not been particularly high in the past. In the first Energy Policy Document, the former minister of Economic Affairs noted that 'Solar cells cannot be considered for application in our electricity production system at the current technical level' (EZ, 1974, p. 159). In 1980 the report 'Future energy situation in the Netherlands' (VDEN, 1980) stated that it seemed unlikely that PV would make a significant contribution to the energy supply of the Netherlands. According to the National Energy Survey 1987 (ESC, 1987), the contribution of PV to the energy supply in 2010 would be negligible, with no more than 10 MW of installed capacity.[35]

By the end of the 1980s the views were changing. The study 'Sustainable energy, a look at the future' (Krekel et al., 1987) included a scenario that foresaw 83 PJ of PV electricity being generated in 2050 (over 20 per cent of total electricity production in that scenario). Alsema and Turkenburg (1988, p. 77) foresaw a production level of over 20 TWh[36] (72 PJ) within 25–40 years after the start of large-scale introduction (assuming sufficiently low cost). The 1990 policy document on energy conservation stated that PV might not yet represent any more than 2 PJ of the overall Dutch energy supply in 2010, but after that it might 'in principle' become the most important renewable energy option (EZ, 1990). A study conducted on behalf of the Ministry of Economic Affairs (Alsema and van Brummelen, 1992) concluded that, technically speaking, PV had great potential for the Dutch energy supply. In an optimistic scenario, 8 per cent of the Dutch electricity consumption in 2020 might come from PV power (but in a pessimistic scenario only 0.3 per cent).

The Follow-up Policy Document on Energy Conservation (EZ, 1993) contained a programme to achieve approximately 250 MW_p of PV power by 2010, good for a saving of some 2 PJ of fossil fuels per year. In the 'Third energy policy document' (EZ, 1996), the contribution of PV to Dutch energy needs in 2020 was estimated at 10 PJ (less than 0.5 per cent of the total primary energy demand). According to CE calculations (Bergsma et al., 1997), the Dutch PV potential could be much higher: 30 TWh in 2010 (over 100 PJ, one quarter of the electricity need) at a cost of less than NLG 0.30 (€0.14) per kWh (under optimistic assumptions). The Dutch government is still reticent, however. In a study by Economic Affairs, 'Energy and society in 2050' (EZ, 2000b), PV plays a substantial role in only one of the four scenarios described.

As for the prospects of further cost reduction and higher efficiency of PV, the literature presents a generally optimistic picture. A study on behalf of Greenpeace (KPMG, 1999), based on European research, concluded that strictly economies of scale (in the production of PV panels) could already lead to a cost reduction (at the current level of technology) of 60–80 per cent. According to Lysen (2003) the price of complete PV systems (now €4–8 per W_p, equivalent to 20–80 eurocents per kWh) could drop to €2–3 by 2010 and ultimately to roughly €1 per W_p, both as a result of economies of scale and ongoing R&D efforts. ECN arrived in 1995 at production costs for PV in 2020 of NLG 0.12−0.23 (€0.05–0.10) per kWh (EZ, 1996). Lenstra (2000) is less optimistic. In his opinion, the price per kWh of PV electricity could only come down to NLG 0.35 (€0.16), which he considers insufficient to be able to compete with other options that are neutral in climate terms, such as biomass and natural gas with CO_2 storage.

As evidenced by this last example, the key question for the future of PV is the break-even price of PV electricity; in other words, the price at which PV can compete with conventional forms of electricity without depending on subsidies or other specific support. Poponi (2003) calculated break-even prices using various assumptions regarding useful life, discount rate and operating time. At an electricity price of $0.05 per kWh (illustrative for the situation in which PV must compete with conventional middle load generation), he arrived at average break-even prices of $0.88 per W_p for complete PV systems and $0.52 for PV modules only. At an electricity price of $0.15 per kWh (typical for building-integrated applications where excess electricity can be redelivered by the user to the grid at end-user prices), the average break-even price amounts to $3.20 per W_p for PV systems and $1.92 for PV modules. Assuming a learning ratio of 20 per cent (see Section 5.4.1) and an average annual market growth of 30 per cent, these break-even prices could be achieved in 2026 and 2011, respectively (at a market growth of 15 per cent, this would be 2045 and 2017).

It is uncertain how long market penetration growth of PV technology can go on. Poponi (2003) mentions two potential constraints:

- Limitation – with a view to the reliability of supply – of intermittent energy sources (such as sun and wind) in terms of their share in the total electricity supply (estimated at 10–30%).
- Costs of materials: the PV industry now benefits from cheap silicon from the micro-electronics industry (see Section 5.4.1), but as the PV market starts to grow significantly, this source will no longer suffice.

The possibility must furthermore be taken into account that electricity production based on fossil fuels also continues to benefit from learning effects, so that the required break-even price of PV would be even lower. Another relevant factor is that the cost of PV is not only determined by the solar cells but also to a significant extent by the 'Balance of System', meaning the other system components (such as converters). It is uncertain whether the potential cost reduction for Balance of System components is as large as for the PV cells.

Van der Zwaan and Rabl (2003) conclude that the share of PV in the global energy supply could become significant after 2020 as a result of cost reductions from learning curve effects. Until that time PV will have to be supported by policy measures. According to the authors, these can be partially justified by the need to incorporate the external costs of fossil-fuel based energy in relevant prices, even though the gap cannot be expected to be fully closed as a result of such measures.

Which PV technology holds most promise for the future is yet unsure. There is a certain consensus, however, that it will be some form of thin-layer PV. Green (2000) expects that a winner from this group of technologies will present itself sometime around 2010. Lysen (2003) notes that there may be several winners, depending on the specific application field.

5.4.3   Driving Forces and Barriers

This section discusses the principal factors that impact the development and application of PV technology and correspond with the aspects listed in Table 2.1.

Diversity
The *firms* that engage in PV vary significantly, both in terms of size (ranging from small PV businesses to multinationals) and business sector (micro electronics, oil industry, chemicals).

The diversity of PV *technologies* is clearly on the rise. While the market for PV is still dominated by cells based on monocrystalline silicon, new technical options are being developed within PV technology (see Section 5.4.1). Mutual competition will have to determine whether these include specific systems that in due time will be able to measure up to the prevailing forms of electricity generation, also in terms of costs.

PV technology is also characterized by a great diversity in forms of and possibilities for *application*. This may be an indicator for likeliness to succeed.

Innovation

New *combinations* and applications of current technologies have served as an important basis for PV. For example, the rise and development of PV technology cannot be viewed separately from semiconductor technology, as both the idea of using silicon and the possibility of obtaining relatively cheap monocrystalline silicon originated with the semiconductor industry. Furthermore, the idea for the development of thin-film technology came from Elliot Berman, a chemist who had conducted work on photographic films before he joined Esso, where he became involved in PV development (Perlin, 1999).

Serendipity also played a role in the development of PV. Solar cells were originally made from selenium, but they were unable to produce enough electricity for practical use. In 1953, Gerald Pearson, a physicist at Bell Laboratories, while conducting research into the application possibilities of silicon for electronic purposes, made a silicon solar cell without ever intending to do so.[37] His colleagues went on to develop this into cells that found practical use in electrical equipment (Perlin, 1999).

At the time, Berman found no response among venture capitalists to his ideas for thin-film PV: 'They weren't very venturesome and what I had wasn't a venture by their definition' (Perlin, 1999, p. 53). On the other hand, in their search for alternatives for long-term energy supply, large oil companies have turned out to be an important source of *venture capital* in the case of PV.

The space industry was the first important *niche market* for PV cells. A significant contributing factor was obviously the fact that the prevailing energy technology possessed few alternatives for the energy supply of satellites and other space vehicles. Contrary to batteries, PV cells have an almost unlimited lifetime. In addition, the permanent availability of sunlight and the absence of weather influences supported the argument for PV applications in outer space. Still, much resistance had to be overcome to get PV accepted in this role. It was mostly due to the perseverance of one individual, Hans Ziegler of the US Army Signal Corps, that this finally succeeded (Perlin, 1999).

The second large niche market was found at sea. In Japan, PV was already used in the early 1960s for lighthouses and beacons. The US Coast guard took longer to equip its light beacons with PV.[38] In the American offshore industry, however, PV cells quickly replaced batteries as the source of energy for the lighting of oil platforms. PV also started to be used for supplying electricity for cathodic protection of oil and gas extraction equipment in remote land areas. According to a spokesman of Automatic Power, the commercialization of PV would have been delayed by at least ten years had it not been for the oil industry (Perlin, 1999).

Other PV applications in remote land areas were found in the telecommunications sector and the electricity supply to rural areas in developing countries.

Grid-connected applications, in particular on the roofs of houses, started to 'get off the ground' in the 1980s. This led to success, especially in Germany, with its 'thousand roofs programme'. In the Netherlands, where the potential for stand-alone systems was limited because of the high density of the electricity network, PV in built-up areas has also become the most important niche market. By integrating solar panels in walls and roofs, cost savings can also be realized on other building materials.

As we saw earlier, PV plays only a minor role in studies of the *futures* of energy supply. In fact, for the Netherlands, PV was for some time hardly seen as a serious option. It is conceivable that this perceived lack of prospects was (and maybe still is) a factor that has slowed down PV innovation.

The selection environment

Physical restrictions do not appear to play a significant role as a selection factor for PV. Although the turnover yield is theoretically tied to a maximum, the upper limits are hardly ever attained in practice, at least not yet. Furthermore, the theoretically achievable yields become higher as newer generations of PV technology appear (see Section 5.4.1).

The *technological possibilities* – and, more so, the costs – constitute a tough barrier for a breakthrough of PV. As we saw in Section 5.4.2, substantial cost reductions through learning curve effects will be necessary to make PV price competitive.

Geographical factors obviously play a significant role. This relates not only to the availability of sunlight, but also to other specific circumstances that can operate to the benefit of PV, such as the absence of an electricity grid.

Perlin (1999) gives a nice example of the role of *market power* in the pioneering days of PV applications. Automatic Power, the quasi-monopolist supplier of lighting systems for oil platforms in the Gulf of Mexico, bought the prototype of Berman's PV navigation equipment but did nothing with it, as the company feared that it might lose its lucrative trade in batteries. Ultimately, PV broke through in the offshore industry via another company (Tideland). The interest in PV by oil companies was also viewed by some as an attempt to destroy the technology by buying the patents and then doing nothing with them. However, this was probably not the case. In the early 1970s these companies were anxiously searching for alternatives to oil, as it was getting scarcer and more costly, and PV looked as though it might be such an alternative. When oil prices started to fall again in the 1980s, some oil companies left the PV business, while others (such as BP and Shell) held on to it.

Government policy (for example, in the form of R&D subsidies and financial incentives for PV applications) has always played a significant role. As long as the cost of PV remains well above that of conventionally produced electricity, government policy will continue to be a key factor.

Path dependence and lock-in

In view of the modular construction of PV systems, a doubling of power implies a doubling of costs. Their *application* can thus hardly benefit from *economies of scale*. PV is thus especially suited for the distributed supply of electricity. Economies of scale do play a significant role in the *production* of PV systems (see Section 5.4.2).

Co-evolution

Large-scale application of PV in the energy system has implications for other system components, since the supply of sunlight is not constant. In order to achieve a balance between the supply of electricity from PV and the demand, energy will thus have to be stored and possibly even transported (from areas with much sunlight to areas with a high energy demand). In addition, mechanisms and incentives can be investigated so as to create a closer match (in space and time) between the demand for energy and the supply of sunlight.

Bounded rationality

A short *time horizon* does not favour a technology such as PV, which requires large capital investments. PV installations have a long life, low maintenance costs and obviously require no fuel. The costs therefore depend heavily on the discount rate or payback period applied by the investor.

5.4.4 Conclusions

PV technology is marked by a high level of diversity and innovation and of interaction with other technologies. The learning curve is completed fairly quickly, considering the presence of various niche markets. For the time being, however, the costs of PV are still too high to be able to compete with other energy supply options on any significant scale.

Government policy has until now focused both on R&D support and market incentives. However, Dutch incentive policy is more reticent than German policy, and the recent focus on the cost-effectiveness of energy subsidies is to the detriment of PV. To maintain an acceptable pace in traversing the learning curve, more support for PV applications will be needed; for example, by providing stimuli to niche markets (housing construction, aid to developing countries) and through government procurement policy. A stringent CO_2 policy will also be essential (though not sufficient in itself) to enable PV to acquire a significant slice of the energy supply pie.

5.5 IMPLICATIONS OF THE CASE STUDY FINDINGS

Several conclusions can be drawn from the case studies in this chapter, which may well also apply to other situations when it comes to the implications of insights from evolutionary economics for the development and application of new energy technology.

In the first place, it should be noted that government policy has mainly focused on providing incentives for R&D. This has contributed to the significant progress that has been made in the realization of better, cleaner and more efficient technologies. However, in order to ensure that the technologies that have been developed also progress in terms of their learning curve, policy measures will also need to focus more on application and diffusion. After all, the independent variable of the learning curve is the cumulative production or application of the technology, and this variable must increase exponentially for an exponential reduction of cost or price to be achieved. When analysing learning curves, we should also not forget that traditional technology can also continue to benefit from learning-curve effects. In other words, the ultimate target is not a cost level that can compete with the current price of the old technology, but with that of its future price, which may be lower as well.

Market stimuli are also essential to prevent chicken-and-egg situations from occurring. Companies and consumers often have a wait-and-see attitude and do not invest in new technology, because they first want to see price reductions and technical improvements: however, these do not come about sufficiently because of the lack of investments in new technology. This deadlock can be broken in various ways, but especially by creating a favourable selection environment (for example, by creating or stimulating niche markets).

It is often feared that stimulation of specific technologies comes down to picking the winners (a practice that has led to poor results and the thought that the market would do a better job at this than the government). It should be made clear, however, that what is especially needed is keeping the playing field wide, promoting diversity, preventing lock-in and creating fair chances.

A problem in creating a favourable selection environment is the fact that the public sector is actually pulling out of areas in which it could create niche markets (such as the energy sector, housing construction and public transport). It would make sense for the government to reconsider its role in these sectors. After all, we are talking about sectors that involve large economies of scale and long-term investments, with the corresponding risks of lock-in and less than optimal results if everything is left to the market.

A potential dilemma facing policy makers with regard to large-scale technological changes is the question of whether or not to invest in improvement of existing technologies and the development of hybrid technologies. On the one hand, this can contribute to lock-in of the old system. On the other, experience tells us that a hybrid technology can mean an advance towards a technological breakthrough. An unambiguous solution to this dilemma cannot be given on the basis of the cases presented here.

It was asserted in Section 2.4.3 that future scenarios (including timetables) could fulfil a useful function as a source of inspiration. The three energy technologies discussed in the case studies still lack coherent and widely supported visions. However, the two recent EU reports of the High Level Group (on fuel cells) and PV-TRAC (on PV) do represent a start in this direction.

The importance of interaction, co-operation and cross-fertilization should also be pointed out. The case studies confirm the view that technical breakthroughs often occur when insights from a totally different business sector or field of study are applied. Policy makers might therefore be advised to bring people together from sectors that are totally unrelated, so that they can stimulate each other with new ideas.

Table 5.4 summarizes a number of characteristics of the three energy technologies considered, including the aspects from evolutionary economics in Table 2.1.

Table 5.4   The characteristics of the three energy technologies and related policies

	Fuel cells	Nuclear fusion	PV
Diversity	High (applications, types, fuels)	Limited	High (applications, types)
Innovation	Strong *interaction* between different sectors (incl. chemicals, energy companies, auto producers)	Expertise concentrated within a small global village; much internal *co-operation* but little external interaction	*Serendipity* and cross-fertilization were keys to development (e.g. thin-film technology)
	Niche markets: aerospace, vehicles	No *niche markets*	*Niche markets*: aerospace, off-grid applications
Cumulative R&D in IEA countries, 1974–98 (in billion US$)	*Unknown*	*26.8*	*5.2*
Selection environment	Liberalization can present opportunities (decentralization); environmental policy is the key (e.g. zero-emission laws)	Not mature enough for market introduction. Viability will depend partly on stringent CO_2 policy	Market power has played a role; government policy (esp. subsidies) is an important factor
Bounded rationality	Application calls for break with existing routines; imitation behaviour in auto industry	Private investors not interested due to long time horizon. Not possible to build on existing routines	Technology involves high capital investments; long time horizon
Path dependence and lock-in	Limited economies of scale in applications:	Heavy path dependence, economies of scale very important:	Limited economies of scale in applications:

Table 5.4   (continued)

	Fuel cells	Nuclear fusion	PV
	→ fits well into decentralized systems *Lock-in* in existing technology (e.g. combustion engine) is a significant barrier	→ difficult to fit into decentralized systems	→ fits well into decentralized systems
Co-evolution	Interaction with other components of energy system (incl. fuel infrastructure) is important	Little interaction with other energy technologies; internal complementarity exists (e.g. plasma physics and materials science)	Implications for other components of the energy system (e.g. due to fluctuations in supply of sunlight)

6. Summary and conclusions

6.1   BRIEF SUMMARY

The aim of this book has been to argue and illustrate that much can be learned from the insights and reasoning provided by modern evolutionary economics for the design of policies and institutions to foster a transition to an environmentally sustainable economy, with a special focus on the transition to a sustainable energy system. For this purpose, a theoretical framework was developed based on six key concepts: diversity, innovation, selection environment, bounded rationality, path dependence and lock-in, and co-evolution. The framework has been derived from a general analysis of the evolutionary economics literature. By applying this framework, insights regarding the content of transition management have been obtained. From this it follows that transition policy is a combination of elements from environmental regulation, stimulating unlocking and innovation policy.

Three types of analyses have been offered to support our findings. First, transition policy and management was analysed at a very general and conceptual level, resulting in a set of guidelines. Second, current and intended Dutch government policies aimed at stimulating technological and organizational innovations in energy systems have been evaluated on the basis of the evolutionary-economic framework, making use of the most important policy documents available. Finally, based on the theoretical framework, a detailed illustrative analysis of the development of three quite different energy technologies – namely fuel cells, nuclear fusion and solar cells – has been carried out. The following sections present a summary of the results and the main policy conclusions.

6.2   THE POLICY FRAMEWORK OFFERED BY EVOLUTIONARY ECONOMICS

Evolutionary economics is nowadays widely regarded as an approach that lends itself well to an understanding of structural change processes related to technology, organizations, economic structures and institutions. The uniformly rational and optimizing behaviour of representative individuals and groups, which has been assumed in traditional economic theory, is replaced in evolutionary economics by a more realistic approach – supported by a wide range of experimental and empirical studies – that is characterized by bounded rationality and diversity of all actors. This bounded rationality takes the form of routines, habits, imitation of others and myopia (i.e. employing a limited time horizon).

Bounded rationality goes hand in hand with heterogeneity in the behaviour of actors, as expressed in the great diversity of behavioural strategies. Therefore, populations occupy a central role in evolutionary economics. Diversity is characterized by three dimensions: variety, balance and disparity. Each of these influences the effectiveness and outcomes of innovation and selection processes. The diversity of elements – which can represent agents, strategies, behaviours, techniques, products, institutions and so on – within a population is influenced by innovation and selection processes. Innovation processes lead to greater diversity, while selection processes have the opposite effect. Selection encompasses many different factors and may be related to physical, technological, geographical, internal business-related, market-related and institutional dimensions. Together they determine the relative suitability of a specific technical, organizational or institutional alternative. Serendipity plays a significant role in innovation, meaning that a combination of coincidence and knowledge can lead to new insights. Major innovations often result from combining existing elements within a population. In other words, diversity itself increases the potential for innovation. Knowledge is essential, since creative innovations nearly always arise from the combination of existing knowledge, techniques (i.e. the aggregate of cumulative knowledge) and concepts.

The dynamics of evolutionary systems results in path dependence and co-evolution. Path dependence means that as a result of increasing returns to scale a self-reinforcing feedback mechanism is put into motion, leading to the dominance of a specific – possibly undesirable – technology or economic structure. The result is a historical irreversible path in terms of the distribution of activities or technologies in a population, which is at the same time extremely sensitive to initial random events. Increasing returns arise on the demand and supply sides of the market, and involve economies of scale, learning effects, network externalities, imitation, information externalities and technological complementarities. Path dependence often ends in a lock-in situation, with a single technology being left. The self-reinforcing returns to scale of this technology make it difficult to escape from such a situation.

Co-evolution is a concept that refers to the mutual impact of distinct evolutionary processes within a system or between systems. This makes it a useful concept for considering complex processes that involve subsystems.

Evolutionary changes in one subsystem are subject to selection pressure from the other subsystem, and vice versa. Co-evolution is different from co-dynamics, which is non-evolutionary dynamic interaction between subsystems. Co-evolution draws attention to the way in which subsystems change over time in a complementary and interactive way, and thus become better matched to each other as well as to the total selection environment.

The foregoing discussion shows that a number of key elements can be distilled from evolutionary economics: diversity, innovation, selection, bounded rationality, path dependence and lock-in, and co-evolution. All these can be subdivided further, as is shown in Chapter 2 (summarized in Table 2.1).

6.3   IMPLICATIONS FOR ENVIRONMENTAL POLICY AND TRANSITION MANAGEMENT

Economic principles play a major role in the shaping of environmental policy. The criterion of efficiency plays a dominant role. For many aspects of environmental policy this makes good sense, but it is less suited for policy aimed at innovations and transitions (major system innovations). Evolutionary theory and related concepts can provide useful insights into the contents of environmental policy. This is particularly relevant when seen against the background of the aim to achieve sustainable development and transition management.

Path dependence serves as an important starting point for this, considering that early and partially haphazard development phases have an irreversible impact on the later developments of a system ('historical accidents'). Great care in the early phase of innovation or transition policy is therefore essential. From an evolutionary perspective, ensuring diversity is a key starting point in designing policies, especially when focused on transitions to sustainability. Diversity ties in well with the broader approach of sustainable development, which is strongly focused on maintaining biodiversity and resilience. In line with this, innovation and transition policies should, especially in early phases, be directed at increasing the diversity of technologies and business strategies.

A level playing field is a situation in which parties and technologies compete with others from similar starting positions. In general, the 'free market' is regarded as the system that ensures maximum competitiveness. The evolutionary-economic approach argues that a level playing field must be considered in a broader, extended sense than strictly as the presence of a free, competitive market. Instead, it emphasizes unlocking of existing structures and prevention of early lock-in. Several other features are important for such an *extended level playing field*. First, prices should reflect all relevant social costs, private as well as external. Next, policy makers must realize that technologies that may be attractive from the viewpoint of sustainability, but that are still at an early stage on the learning curve, have an unfavourable competitive position compared to technologies that are further along the learning curve. In this situation, free market competition is not always beneficial from the perspective of sustainability and transition management. It is also important to avoid an unwanted early lock-in of specific technologies. This may require that at the start of a transition process, the (increasing) returns to scale that favour one technology over the other are constantly corrected. Lastly, differences in selection environment (other than markets) should be corrected so that alternative options are exposed to comparable selection environments, leading to fair competition. The result of the consequent extended level playing field in an evolutionary sense is that in the long term a greater diversity of technologies will arise, so that the chance of a single unwanted technology getting locked-in is reduced.

To summarize, the creation of an extended level playing field implies a broader approach than traditional economic policy. It requires not only a good functioning of the market, but also setting the conditions for the market to perform within the boundaries of sustainability, to stimulate useful innovation trajectories, and to be consistent with unlocking and avoiding early lock-in.

The concept of lock-in also plays a role at a more subtle level. A policy that concentrates on energy conservation, clean fossil fuels (CO_2 capture and storage) or combined heat and power (CHP) could lead to reinforcement of fossil fuel lock-in, so that a transition to sustainable, renewable energy sources might be delayed. In view of this, extra efforts or investments directed at energy conservation ought to be combined with extra stimuli to generate renewable energy. Otherwise it will be increasingly difficult to undo the lock-in of fossil fuels. It may also be necessary to subject the stimulation of energy conservation to specified preconditions that do not favour fossil fuels or that do not discriminate against renewable energy sources.

Diversity, the central notion of evolutionary thinking, is a multidimensional concept that is characterized by variation, balance and disparity. Variation refers to the different alternatives (technologies, processes, products, organizations, institutions or strategies) that are encountered in a population of elements. Balance relates to the extent to which one or more elements dominate in a population, either in terms of size or number. Disparity indicates the degree of difference between elements in a population. Each of these three dimensions of diversity influences the outcomes of innovation and selection. Policy therefore ought to pay careful attention to each of the three diversity dimensions.

Radical innovations often arise from creative combinations, which can benefit from the presence of much diversity (variation as well as disparity) in technologies and strategies. With this in mind, wastage may well be fostered; after all, without wastage there would be no major innovations. Pursuing this line of thinking, unusual combinations of knowledge, concepts and technologies may open doors to worthwhile innovations. Governments can play a role here, for example, by stimulating the creation of networks and bringing experts from the natural and social sciences together, with the aim of mutual inspiration and multidisciplinary solutions to prevailing problems and barriers. From a co-evolutionary perspective it further makes sense to stimulate a continuous dynamism of potentially creative combinations.

A certain degree of spatial or economic isolation of experiments and innovation initiatives should be fostered, as it can lead to unique pathways. The reason is that isolation allows new, young alternatives (technologies, ideas) to stay outside the influence of the dominant technology and ways of thinking (paradigm). Niche markets are a practical example of economic isolation. In addition, allowing or even supporting a diversity of initiatives at a local level (municipalities and provinces) can contribute to stimulation of more or less isolated experimental sites and breeding grounds.

From an evolutionary-economic perspective, government policy should thus not focus on picking the winners but on creating a suitable innovation and selection environment, which allows for certain options to perform better than others. While some dimensions of the selection environment (the physical and geographical dimensions) are hardly subject to government intervention, others may well be sensitive to public policy. For each innovation target, the selection environment must be carefully identified, including its dynamics in terms of time.

It is important to realize that evolutionary economics does not necessarily lead to 'optimal policy' in the traditional sense. In neoclassical economics it is assumed that companies and consumers demonstrate optimizing behaviour. In evolutionary economics, on the other hand, pricing instruments are regarded as generally insufficient to realize environmental policy objectives. While these instruments influence the choices made by actors, such choices are characterized by bounded rationality – in the form of routines, habits, imitation or myopia. The effectiveness and efficiency of pricing instruments is consequently overestimated by the traditional economic theory of environmental policy. These imperfections are reinforced by the effect of lock-in. As a result, inclusion of external environmental costs in prices will be insufficient – even if necessary – to achieve a transition to sustainable development.

Within an evolutionary type of economic policy it is crucial to find a proper balance between efficiency (or cost-effectiveness) and diversity. This can be translated into a trade-off between the costs of diversification in the short term – due to lack of economies of scale – versus the benefits of diversification in the long term – due to innovation and selection opportunities. Given large uncertainties about the future, this trade-off can never be made fully objective (for example, through a cost–benefit analysis). Instead, it will take the form of a qualitative assessment on the basis of the opinions and intuitions of a broad range of experts and stakeholders. On a larger and international scale, the consideration or balance between diversity and scope is less difficult, since economies of scale and other (increasing) returns to scale can be significant even when a significant amount of variation in technologies or institutions continues to exist. For policy-setting purposes this represents the challenge of achieving a proper balance, through collaboration between countries, of the investments in sustainable innovation projects, where scale and diversity serve as major criteria.

Even though evolutionary economics views the economic system fundamentally as an autonomous process without any particular objective, it is clear from the above outline that normative elements (aimed, for example, at sustainability) can be added to this system by means of policy. This is done through policy measures that directly impact the innovation and selection environment. It results in a partly controlled evolutionary process that allows a transition to sustainable development.

6.4   AN EVOLUTIONARY-ECONOMIC EVALUATION OF CURRENT DUTCH ENERGY INNOVATION POLICIES

The key concepts of evolutionary economics (diversity, innovation, selection environment, bounded rationality, path dependence and lock-in, and co-evolution) can be applied to evaluate current policies aimed at stimulating energy innovations. Environmental policy, transition policy, energy policy and innovation policy all play roles in this. An important development since the start of the present century has been the adoption of the transition approach, which in a theoretical sense can be seen as moving from a more or less linear concept of technology development to a full systems approach. In innovation policy this approach was introduced some years earlier, but only with the purpose of stimulating the development of innovations, rather than 'innovating' the system itself. This development was made operational by fostering networks of companies, technology institutes and research institutes, as well as making the development, transfer and use of knowledge a specific objective of policy. The focus of policy instruments, however, has continued to be on traditional regulatory and price instruments. System instruments, such as innovation networks and integrated innovation programmes, are not yet being widely applied. Nevertheless, interest in such system instruments seems to be increasing, notably due to the impact of transition policy.

Many of the key concepts of evolutionary economics are recognizable in current policy aimed at energy innovations, but their follow-up into concrete policy is often limited. For example, while the importance of maintaining diversity is recognized in policy documents, this is mainly applied to technologies and less to companies, products and strategies. There is debate in policy about the dilemma between the creation of sufficient mass (to prevent dissipation of budget funds) and the maintenance of diversity.

The concept of innovation is frequently discussed, but from the perspective of evolutionary economics it is approached in a rather restricted way. Attention to co-operation and future vision scores highly, but in the reviewed policy documents the development of schooling and venture capital seldom comes up. Aspects such as cross-fertilization, serendipity, isolation and niche markets, from an evolutionary-economic perspective all judged to be quite important for innovation and the innovation climate, hardly receive any attention in policy. In some cases policy documents even evaluate these aspects in a negative way. Presently, energy innovation policy still focuses more on the development of innovations via R&D than on diffusion and application. In this respect, the more system-oriented elements of evolutionary economics are not all equally reflected in public policy. The role of the government as a launching customer will also need to be further elaborated.

Considerable attention is paid to the 'hard side' of technological development, whereas the other aspects (social embedding, institutions) tend to get overlooked. This may be explained by recognizing that the government too is somewhat bounded in its rationality. Policy development is myopic when it comes to stimulating technologies that are at an early stage on the learning curve. The European level playing field receives much attention in the policy documents, motivated by the idea that an open European market determines the playing field for the Dutch government to a considerable extent. This means that industry, business sectors and technologies that cannot compete economically in this market should not expect any government support. An approach that is more inspired by evolutionary economics would, on the other hand, place much greater emphasis on the underdevelopment of certain technologies, on the lack of pricing of external effects (environmental pressure) caused by old technologies, and on stimulating a broad spectrum of technologies for the long term.

The concept of selection environment scores poorly in current policy. The predominant principle in innovation policy is that 'the market must choose'. Public authorities do not seem to recognize that the selection environment contains other factors, which can be influenced by public policy. A mainly negative perspective arises from the policy documents; namely, that the 'government disrupts the operation of the market'. Much attention is devoted to removing barriers to innovation, while opportunities to change the selection environment in favour of innovations serving the public good are rare.

The concept of bounded rationality scores well in policy documents, in particular with regard to the time horizon of entrepreneurs and imitation behaviour. It should be noted, however, that governments themselves often act in a routine way, which is reflected by public decisions depending on a limited and incomplete set of policy insights.

The notions of path dependence, lock-in and the level playing field have also found their way into government policy. The elaboration of this into concrete policy is, however, rather one-sided from an evolutionary-economic perspective. Government authorities appear to want to prevent lock-in mainly by postponing selection instead of by conscious stimulation of flexible options. Especially with regard to energy policy, a debate about flexible options in the context of sustainability would be appropriate, for example, in deciding between large-scale and small-scale generation. The aspect of the level playing field is strictly viewed as the creation of equal competitive positions among producers from different countries. An extended playing field as discussed in the previous section, involving technologies that are in different phases of the learning curve, is altogether ignored in the discussion. Coherent and widely supported scenarios related to specific energy technologies could function as a source of inspiration and as a binding element between the different players in the community setting and, in this way, realize support for and stimulate creative innovations.

The assessment of Dutch policy on energy innovation shows that some elements of evolutionary economics are implicit, few are explicit and most are entirely lacking. It seems that it is mainly those evolutionary-economic concepts that do not lead to strong tensions with the policy suggestions that arise from traditional economic theory that have found a place in policy documents. Many of these correspond with more generic developments in innovation policy, supported by changes in theory from a linear model for technology management (in the 1970s) to cluster policy (in the early 1990s) and from there to present-day management of the total innovation system. This growing complexity is especially visible in transition policy. The role of policy corresponding to a hierarchical style of government is gradually giving way to a more network-oriented approach. The government's role will change then from 'overseer' to manager.

6.5   EVOLUTIONARY CONCEPTS APPLIED TO SPECIFIC ENERGY TECHNOLOGIES

In the context of this study, the evolutionary-economic framework has also been applied to three specific technologies; namely, fuel cells, nuclear fusion, and photovoltaic cells.

Fuel cells are characterized by a large degree of variability in companies, technologies and potential applications. With regard to this technology, the Netherlands as well as the EU are still strongly focusing their attention on the R&D phase. The learning curve up to and including the possibility of commercial application is still steep. The application and diffusion of this technology will now have to be stimulated. This calls for devoting attention to niche markets, hybrid applications and the role of the government as a launching customer. The application of fuel cells will ultimately require a break with current routines, especially those associated with fossil fuels and the combustion engine.

The road to commercial application for *nuclear fusion* is still extremely long, and the related research is mainly fundamental in nature. As a result, the business community is as yet hardly involved in this technological development. The learning curve is nonetheless developing quite fast. Since the economies of scale are huge, distributed and small-scale application is out of the question. The diversity of technologies and parties is very limited. Nevertheless, the prospect for the future is very promising, since nuclear fusion as a source of energy is potentially cheap and causes hardly any pollution. Public investments in research are justified primarily because of the short time horizon of private investors.

Photovoltaic cells (PV), or solar cells, are the opposite of nuclear fusion in the sense that they involve a distributed form of energy generation. The silicon solar cell was invented more or less by accident in the electronics industry (a case of serendipity), and the idea for a thin-film solar cell came from photography (cross-fertilization). Niche markets for photovoltaic cells have come about especially in the space industry, and in applications for very remote areas, such as marine light beacons. PV can be linked to the electricity grid but can also operate independently. The opportunities for application are thus fairly substantial, but so are the investment costs, despite the rapid movement along the learning curve. PV benefits little from a large scale of application. In the Netherlands the potential of PV is limited because of the dense electricity grid. Distributed applications are thus mostly confined to niche markets. Large-scale applications require a total break with the existing system of energy supply.

6.6   MAIN POLICY CONCLUSIONS

The evolutionary-economic framework, as elaborated in this book and applied to Dutch energy innovation policies as well as to a number of important energy technologies, yields some important conclusions for policy makers.

The key message of applying evolutionary economics to environmental policy and transition management is to find a proper balance between efficiency and diversity, or between the short-term costs and long-term benefits of diversification. This is of course not a simple aim. As innovation and the effects of selection inevitably involve uncertainty, finding the balance cannot be treated as a traditional question of weighing costs versus benefits. More research will be needed to resolve this issue, notably with the aim of developing suitable methods that can be made operational. A likely strategy in the meantime could be to uphold the following heuristic: the greater the uncertainty, the more diversity needs to be maintained or stimulated. At an international level it is easier to find a balance between efficiency and diversity because of the much larger scale. This provides a strong case for more co-operation and alignment of policy, innovation programmes and investments between different countries.

An extended level playing field is needed to bring about sustainable innovations and transitions. In addition to the traditional features of open market competition, this will have to include the following: (1) prices that reflect external costs; (2) stimulation of technologies that are attractive from the point of view of sustainability and that are at a relatively early stage on the learning curve; (3) appropriate adjustment of the increasing returns to scale of an alternative so as to prevent premature lock-in; and (4) a selection environment that should be levelled as much as possible.

Major innovations and a transition to sustainable development are never without wastage or dead ends. Variability at all levels and in all dimensions (processes, products, organizations, institutions and space) must be fostered. Fisher's theorem reasons as follows: 'The greater the genetic variability upon which selection for fitness may act, the greater the expected improvement in fitness.' Furthermore, a higher level of variability supports major innovations, since it allows more combinations of existing concepts or technologies. With this in mind, it is relevant to stimulate interaction between economic sectors, since this increases the likelihood of creative innovations through combinations, mutual exchange, co-operation and cross-fertilization. Coherent and widely supported visions with regard to specific energy technologies can function in this respect as sources of inspiration and as binding elements between the different players in the community at large, thereby stimulating creative innovation. Because of the positive effects of variability both in innovation and selection, it is unwise to apply tight planning and early selection of winners in transition policy, as this would significantly lessen the chance of variability. Policy aimed at picking the winners does not in general offer much perspective, since it is not known in advance who or what the winners will be in the long term. It is more effective to keep the playing field wide so as to stimulate variability, avoid lock-in and create equal chances.

As a closing remark, it is good to emphasize once more that the view of evolution as an autonomous process without purpose has been often misinterpreted as implying a cynical perspective in terms of irrelevance and futility of policy. But the inherent aimlessness of evolution – an analogue to the inherent aimlessness of formal markets – does not imply that public policy is ineffective in providing direction to economic evolution or in stimulating the pace of innovations. Indeed, evolutionary development can be guided by a combination of various policies, such as environmental regulation, innovation policy, industrial policy and unlocking policy. In this way, the two dynamic elements of evolution, selection and innovation, will be appropriately affected. Under evolutionary conditions, such a set of policies is incapable of realizing very specific outcomes. This should not be seen as problematic. In the first place, evolutionary progress is more relevant than realizing predetermined or preconceived outcomes. Moreover, traditional policies that neglect evolutionary mechanisms will generally not perform better, and often worse, in terms of evolutionary progress and long-term sustainable human welfare. In other words, in the face of economic evolution as a reality, one can be optimistic about the effectiveness of policies aimed at fostering a transition to a sustainable future.

Notes

1. For a detailed description of these policy areas, see Chapter 4.
2. Successive policy documents continued to foresee growth in the number of nuclear power plants in the Netherlands, even after the 'Broad Social Debate', which was initiated in 1979, made clear that the majority of the Dutch population was against nuclear power. The disaster at the Chernobyl nuclear power plant in 1986 led to the Dutch cabinet no longer seeing further development of nuclear energy as a viable option. However, in NEPP-4 (VROM, 2001) nuclear energy was once again mentioned as an option, mainly in light of the climate issue.
3. For an overview, see the website of the Ministry of Economic Affairs: www. energietransitie.nl
4. This section draws heavily on Kern (2000).
5. An early proponent of the innovation system concept is Michael Porter (1990). The concept of the innovation system was developed further by Nelson (1993), Lundvall (1992) and Edquist (1997), among others.
6. Note that the concept of research and innovations as a driving force for economic growth very much resembles the approach of the 1960s and 1970s.
7. On the other hand, it is also conceivable that investments shift to power plants fired on relatively cheap coal, which is, if no preventive measures are taken, more polluting than existing fossil options such as natural gas. This shift can presently be seen with some investors in the Netherlands, who are clearly taking a short time horizon perspective.
8. This reaction takes place indirectly (via an electrolyte) and with the aid of a catalyst (such as platinum). The exact process depends on the type of electrolyte. In a Proton Exchange Membrane fuel cell (PEMFC), H^+ ions travel via the electrolyte (a membrane) from the anode to the cathode side of the fuel cell, while the electrons that are released through the catalyst traverse a loop via the external electric circuit, where they can perform a useful function. On the cathode side, H^+ ions, electrons and O^{2-} ions come together to form water.
9. There were also high expectations from military applications in the USA. During the Vietnam War, fuel cells were used as a portable energy supply for all types of equipment. Problems were encountered, however, due to leakage and the weight involved (Schaeffer, 1998, p. 362).
10. By comparison, a conventional electric power plant has an average capacity of several hundred megawatts.
11. Hydrogen can also be obtained from fossil fuels by means of 'reforming'. A process has also recently been developed whereby natural gas is converted in the fuel cell itself into hydrogen. The drawback of this from an environmental viewpoint is that CO_2 is still emitted, unless it is collected and stored (below ground or sea).
12. The idea that the natural gas network would obviously also be suited for the transport of hydrogen has subsequently become subject to discussion (see, e.g., Hoffmann, 2001, pp. 198–9). Recently, a project called 'Naturalhy' was started (under the leadership of Gasunie Research), which examines the possibilities of using existing natural gas networks for mixtures of natural gas and hydrogen. See http://www. gasunieresearch.nl
13. Nonetheless, the research conducted by Broers and Ketelaar had a significant impact and formed the basis for research in France and the USA (Schaeffer, 1998, p. 353).
14. TNO has, however, been involved in more recent fuel cell projects (van der Hoeven, 2001).

15. The reason for the initial enthusiasm of steel producer Hoogovens was that it wanted to make more efficient use of its hydrogenous coal gas. Ultimately the company decided to withdraw from the programme because of the high yield requirements applied to all activities not directly related to steel production. Based on an American consultancy report, Hoogovens concluded that the MCFC payback period would be too long and would involve too much uncertainty in the long term (van der Hoeven, 2001).

16. In view of the currency fluctuations between the euro and the US dollar in the past few years (where the rate of the euro has been below as well as above $1.00), we have decided not to translate dollars to euros and vice versa in this chapter. Both currencies are therefore used, depending on the source.

17. By comparison, the investment costs of a STEG power plant (steam and gas turbine) heated with natural gas amount to nearly €600 per kW. Those for a coal dust plant are €1200 per kW, for a wind turbine on land €1000 per kW and for a wind turbine at sea €2000 per kW (Menkveld, 2004).

18. According to Schaeffer (1998), a fuel cell engine would have to cost less than $60 per kW to warrant its introduction in passenger cars, and the fuel cell itself less than $30 per kW.

19. Article in *Stromen*, p. 2: see 'No future for fuel cell' (Anon., 2004).

20. Computer producers such as NEC, Hitachi and Toshiba are planning to introduce laptops with fuel cells on to the market within a few years (see Bakker, 2004, p. 14).

21. The nuclear fusion of deuterium atoms from a mere one litre of water is so great that it corresponds to more than 100 litres of petrol (Fast, 1980). Deuterium is a hydrogen isotope that is used in nuclear fusion reactors (during the same process, lithium is converted into tritium, another hydrogen isotope). Regular hydrogen is not suitable for controlled nuclear fusion under terrestrial conditions.

22. An alternative technique consists of heating very small, free-falling deuterium–tritium balls using laser light (Lysen, 1977). Current research, however, concentrates on tokamak technology and other methods of plasma containment, such as the *stellerator* and inertial containment (see Westra, 2004).

23. Calculated on the basis of IEA figures (in prices and exchange rates of 2002).

24. The head of nuclear fusion research at Rijnhuizen, Professor Dr Niek Lopes Cardozo, received the Royal Dutch Shell award in 2003. He is not only engaged in research but also seeks to interest the public at large (and especially secondary school students) in nuclear fusion through an interactive 'Fusion Road Show'.

25. As expressed by Minister Jorritsma during general consultations with the Lower House regarding the Energy Research Strategy (EOS) on 6 December 2001 (Parliamentary Document 28 108, no. 2). Research by 'Rijnhuizen', however, is partly financed by Dutch budgets for science (via FOM and NWO).

26. A programme was nonetheless started up in the 1970s in the USA with the intention of having an economically feasible nuclear fusion power plant in operation before the end of the 20th century (LSEO, 1975).

27. Westra (2004).

28. The maximum deliverable capacity of PV installations is expressed in megawatt peak (MW_p).

29. Esso (today's Exxon) introduced PV modules on to the market as early as 1973 via its subsidiary Solar Power Corporation (Perlin, 1999).

30. Source: press release by Shell Solar, 25 October 2002.

31. The World Bank, which originally saw no future in PV, now has a separate department that employs a large number of specialists who work on PV projects in developing countries (Perlin, 1999, pp. 188–9).

32. Other publications (such as Poponi, 2003) use the term 'progress ratio', referring to the value 2^b (given the symbols applied here).

33. The final report of the PHOTEX project is available at http://www.energytransition.info/photex/

34. Calculated on the basis of IEA figures (in prices and exchange rates of 2002).

35. As Figure 5.5 shows, the installed capacity barrier of 10 MW was already surpassed in 2000.

36. The report erroneously mentions 20 GWh.
37. However, according to Riordan and Hoddeson (1997, quoted in Green, 2000), it was Russel Ohl who happened to discover the first silicon solar cell in 1940.
38. In the Netherlands too, light buoys and beacons were the first niche market of any size (see Alsema and Turkenburg, 1988, p. 76).

References

Aghion, P., and P. Howitt (1998). *Endogenous Growth Theory*. The MIT Press, Cambridge, MA.

Alchian, A. (1950). Uncertainty, evolution and economic theory. *Journal of Political Economy* **58**, 211–22.

Alsema, E.A., and M. van Brummelen (1992). Minder CO_2 door PV. Studie naar de maximaal haalbare energie-opwekking en CO_2-emissiereduktie met behulp van zonnecelsystemen in Nederland tot 2020. Ministry of Economic Affairs, Beleidsstudies Energie 3, The Hague.

Alsema, E.A., and W. Turkenburg (1988). Elektriciteit in Nederland met zonnecellen. Publikatiereeks Milieubeheer (1), Ministry of Housing, Spatial Planning and the Environment (VROM), March 1988.

Anon (2004). Brandstofcel geen toekomst. *Stromen*, 13 February 2004, 2.

Arthur, B. (1989). Competing technologies, increasing returns, and lock-in by historical events. *Economic Journal* **99**, 116–31.

AWT (2003). Netwerken met kennis, kennisabsorptie en kennisbenutting door bedrijven. Adviesraad voor Wetenschap en Technologie (advies 56), The Hague.

AWT (2004). Tijd om te oogsten, vernieuwing in het innovatiebeleid. Adviesraad voor Wetenschap en Technologie (advies 59), The Hague.

Ayres, R.U. (1994). *Information, Entropy and Progress: Economics and Evolutionary Change*. AIP Press, American Institute of Physics, New York.

Bäck, Th. (1996). *Evolutionary Algorithms in Theory and Practice: Evolution Strategies, Evolutionary Programming, Genetic Algorithms*. Oxford University Press, New York.

Bakker, J. (2004), Handcomputer of brandstof. *Computable* **19**, 7 May 2004.

Baumol, W.J. and W.E. Oates (1988). *The Theory of Environmental Policy*, 2nd edn. Cambridge University Press, Cambridge, UK.

Bauquis, P.-R. (2003). Reappraisal of energy supply–demand in 2050 shows big role for fossil fuels, nuclear but not for nonnuclear renewables. *Oil & Gas Journal*, 17 February 2003, 20–9.

Becker, G.S. (1976). Altruism, egoism, and genetic fitness: economics and sociobiology. *Journal of Economic Literature* **14** (3), 817–26.

Beek, A., van, M. Maris, P. Heidbuurt and J. Roersen (2003). National survey report of PV power applications in the Netherlands. International Energy Agency, Co-operative Programme on Photovoltaic Power Systems. BECO Group BV, Zwolle, May–June 2003.

Bergh, J.C.J.M. van den (2004). Evolutionary thinking in environmental economics: retrospect and prospect. In: J. Foster and W. Hölzl (eds), *Applied Evolutionary Economics and Complex Systems*. Edward Elgar, Cheltenham, UK and Northampton, MA, USA, pp. 239–75.

Bergh, J.C.J.M. van den (2005). Evolutionary analysis of economic growth, environmental quality and resource scarcity. In: R.U. Ayres, D. Simpson and M. Toman (eds), *Scarcity and Growth in the New Millennium*. Resources for the Future, Washington, DC, pp. 177–97.

Bergh, J.C.J.M. van den and J.M. Gowdy (2000). Evolutionary theories in environmental and resource economics: approaches and applications. *Environmental and Resource Economics* **17**, 37–57.

Bergh, J.C.J.M. van den and J.M. Gowdy (2003). The microfoundations of macroeconomics: an evolutionary perspective. *Cambridge Journal of Economics* **27** (1), 65–84.

Bergh, J.C.J.M. van den, and S. Stagl (2003). Coevolution of economic behaviour and institutions: towards a theory of institutional change. *Journal of Evolutionary Economics* **13** (3), 289–317.

Bergh, J.C.J.M. van den A. Ferrer-i-Carbonell and G. Munda (2000). Alternative models of individual behaviour and implications for environmental policy. *Ecological Economics* **32** (1), 43–61.

Bergsma, G.C., S.A.H. Moorman, J. Verlinden and F.G.P. Corten (1997). Het potentieel van PV op daken en gevels in Nederland. CE (report 97.3789.002), Delft.

Boschma, R.A., K. Frenken and J.G. Lambooy (2002). *Evolutionaire economie, een inleiding*. Coutinho, Bussum.

Boulding, K.E. (1966). The economics of the coming spaceship earth. In: H. Jarret (ed.), *Environmental Quality in a Growing Economy*. Johns Hopkins University Press, Baltimore, MD, pp. 3–14.

Boulding, K.E. (1978). *Ecodynamics: a New Theory of Societal Evolution*. Sage Publications, Beverly Hills, CA.

Boulding, K.E. (1981). *Evolutionary Economics*. Sage Publications, Beverly Hills, CA.

Boyd, R., and P.J. Richerson (1985). *Culture and Evolutionary Process*. The University of Chicago Press, Chicago.

Braams, C.M. (1982). Kernfusie. In: C.D. Andriesse and A. Heertje (ed.), *Kernenergie in Beweging, Handboek bij Vraagstukken over Kernenergie*. Keesing Boeken, Amsterdam, pp. 134–41.

Butter, F.A.G. den, and M.W. Hofkes (2006). A neo-classical view on technological transitions. In: X. Olsthoorn en A. Wieczorek (eds). *Understanding Industrial Transformation: Views from Different Disciplines*. Springer-Verlag, Berlin, pp. 141–62.

Campbell, N.A. (1996). *Biology*, 4th edn. Benjamin/Cummings, Menlo Park, CA.

Comyns, A. (2004). *The Fuel Cells Industry Worldwide: a Market/Technology Report*. Materials Technology Publications, Watford, UK.

Conlisk, J. (1989). An aggregate model of technical change. *Quarterly Journal of Economics* **104**, 787–821.

CPB, Rathenau Insituut, RIVM-MNP, RPB and SCP (2003). Investeren in kennis, een maatschappelijk-economische beoordeling van de BSIK-projecten. CPB, The Hague.

Criqui, P., and N. Kouvaritakis (2000). World energy projections to 2030. *International Journal of Global Energy Issues* **14** (1–4), 116–36.

Dennett, D. (1995). *Darwin's Dangerous Idea: Evolution and the Meanings of Life*. Simon and Schuster, New York.

Diamond, J. (1997). *Guns, Germs and Steel: the Fates of Human Societies*, W.W. Norton, New York.

Dietz, F.J., and H.R.J. Vollebergh (1999). Explaining instrument choice in environmental policies. In: J.C.J.M. van den Bergh (ed.), *Handbook of Environmental and Resource Economics*. Edward Elgar, Cheltenham, UK and Northampton, MA, USA, pp. 339–51.

Dosi, G., C. Freeman, R. Nelson, G. Silverberg and L. Soete (eds.) (1988). *Technical Change and Economic Theory*. Pinter, London.

Dunn, S. (2002). Hydrogen futures: toward a sustainable energy system. *International Journal of Hydrogen Energy* **27**, 235–64.

Durham, W.H. (1991). *Coevolution: Genes, Culture and Human Diversity*. Stanford University Press, Stanford, CA.

ECN (1999). Brandstofcellen, op weg naar de toekomst. In: *Energie Verslag Nederland 1998*. ECN, Petten, Chapter 2, pp. 1–8.

ECN (2004). www.ecn.nl, accessed 18 February 2004.

Edquist, C. (1997). *Systems of Innovation: Technologies, Institutions and Organizations*. Pinter, London.

Ehrlich, P.R., and P.H. Raven (1964). Butterflies and plants: a study in coevolution. *Evolution* **18**, 568–608.

Eldredge, N., and S.J. Gould (1972). Punctuated equilibria: an alternative to phyletic gradualism. In: Schopf, T.J.M. (ed.), *Models in Paleobiology*. Freeman Cooper, San Francisco, CA, pp. 82–115.

Epstein, C., and R. Axtell (1996). *Growing Artificial Societies: Social Science from the Bottom Up*. The MIT Press, Boston.

ESC (1987). Nationale energie verkenningen 1987. Energie Studie Centrum (ESC-42 report), Petten.

European Commission (2000). Presidency Conclusions of the Spring 2000 Lisbon Summit Meeting. http://ue.eu.int/ueDocs/cms_Data/docs/ press Data/en/ec/00100-r 1.en0.htm

European Commission (2003). Hydrogen energy and fuel cells, A vision of our future. Summary report, High Level Group for Hydrogen and Fuel Cells.

European Commission (2004a). A vision for photovoltaic technology for 2030 and beyond. Preliminary report by the Photovoltaic Technology Research Advisory Council (PV-TRAC), presented at the conference 'Future Vision for PV: a Vision for PV Technology for 2030 and Beyond', Brussels, 28 September 2004.

European Commission (2004b). Buying green! A handbook on environmental public procurement. Commission staff working document, SEC (2004) 1050, Brussels.

EZ (1974). Energienota. Tweede Kamer 1974–75 (13 122, no. 1–2), The Hague.

EZ (1979). Nota energiebeleid. Tweede Kamer (15 802, no. 1 and 2), The Hague.

EZ (1990). Nota energiebesparing. Tweede Kamer (21570, no. 1 and 2), The Hague.

EZ (1993). Vervolgnota energiebesparing. Tweede Kamer 1993–94 (23 561, no. 1–2), The Hague.

EZ (1996). Derde energienota. Tweede Kamer (24 525, no. 1–2), The Hague.

EZ (1999). Energy report 1999. Ministry of Economic Affairs (publication no. 13-B-65)/House of Representatives (26 898, no. 1), The Hague.

EZ (2000a). De Kenniseconomie in zicht, De Nederlandse invulling van de Lissabon Agenda voor 2001. Tweede Kamer (27406, no. 2), The Hague.

EZ (2000b). Energie en samenleving in 2050, Nederland in wereldbeelden. Ministry of Economic Affairs, The Hague.

EZ (2001). Energy research strategy (EOS). Ministry of Economic Affairs (publication no. 02ME04), The Hague.

EZ (2002). Investing in energy; choices for the future. Energy report 2002. Ministry of Economic Affairs (publication no. 02ME21), The Hague.

EZ (2003a) Rapport implementatie EOS. Ministry of Economic Affairs, The Hague, February 2003.

EZ (2003b). Action for innovation, tackling the Lisbon ambition. Ministry of Economic Affairs (publication date of English version 9 June 2004), The Hague.

EZ (2004a). Innovatie in het energiebeleid. Energietransitie: stand van zaken en het vervolg. Ministry of Economic Affairs, The Hague.

EZ (2004b). Industry memorandum: heart for industry. Ministry of Economic Affairs (publication date of English version 8 January 2005), The Hague.

EZ and VROM (1991). Technologie en milieu, technologie als schakel tussen ecologie en economie. Ministries of Economic Affairs and of Housing, Spatial Planning and the Environment, The Hague.

Faber, A., and D. van Welie (2004). Onderzoek voor duurzame ontwikkeling, research & development voor transities. RMNO report V.04. Lemma, Utrecht.

Faber, A., T. Rood, and J. Ros (2003). Evaluation of early processes in system innovation, a pilot study on the transformation of Dutch agriculture and food chain to sustainability. Paper presented at conference 'Governance for Industrial Transformation', Berlin, 5–6 December 2003.

Faber, M., and J.L.R. Proops (1990). *Evolution, Time, Production and the Environment*. Springer-Verlag, Heidelberg.

Fagerberg, J. (1988). Why growth rates differ. In: G. Dosi, C. Freeman, R. Nelson, G. Silverberg and L. Soete (eds) (1988), *Technical Change and Economic Theory*. Pinter, London, pp. 432–57.

Fast, J.D. (1980). Energie uit atoomkernen. Natuur en Techniek, Maastricht.

Fehr, E., and U. Fischbacher (2002). Why social preferences matter – the impact of non-selfish motives on competition, cooperation and incentives. *Economic Journal* **112**, C1–33.

Fehr, E., and S. Gächter (1998). Reciprocity and economics: the economic implications of homo reciprocans. *European Economic Review* **42** (3–5), 845–59.

Feldman, M.W., and K.N. Laland (1996). Gene–culture coevolutionary theory. *TREE* **11**, 453–7.

Ferrer-i-Carbonell, A., and J.C.J.M. van den Bergh (2004). A microeconometric analysis of determinants of unsustainable consumption in the Netherlands. *Environmental and Resource Economics* **27**, 367–89.

Fine, G., and J. Deegan (1996). Three principles of serendipity: insight, chance and discovery in qualitative research. *International Journal of Qualitative Studies in Education* **9** (4), 434–47.

Fisher, R.A. (1930). *The Genetical Theory of Natural Selection*. Clarendon Press, Oxford.

Foss, N.J. (1993). Theories of the firm: contractual and competence perspectives. *Journal of Evolutionary Economics* **3**, 127–44.

Foster, J., and J.S. Metcalfe (2001). Modern economic perspectives: an overview. In: J. Foster and J.S. Metcalfe (eds), *Frontiers of Evolutionary Economics; Competition, Self-organisation and Innovation Policy*. Edward Elgar, Cheltenham, UK and Northampton, MA, USA, pp. 1–16.

Freeman, C. (1996). The greening of technology and models of innovation. *Technological Forecasting and Social Change* **53**, 27–39.

Friedman, D. (1998a). On economic applications of evolutionary game theory. *Journal of Evolutionary Economics* **8** (1), 15–43.

Friedman, D. (1998b). Evolutionary economics goes mainstream: A review of the theory of learning in games. *Journal of Evolutionary Economics* **8**, 423–32.

Friedman, M. (1953). On the methodology of positive economics. In: M. Friedman, *Essays in Positive Economics*. The University of Chicago Press, Chicago, pp. 3–43.

Galor, O., and O. Moav (2002). Natural selection and the origin of economic growth. *Quarterly Journal of Economics* **117**, 1133–92.

Geels, F.W. (2002a). Technological transitions as evolutionary reconfiguration processes: a multi-level perspective and a case-study. *Research Policy* **31** (8/9), 1257–74.

Geels, F.W. (2002b). Understanding the dynamics of technological transitions: a co-evolutionary and socio-technical analysis. PhD thesis, Twente University Press, Enschede.

Geels, F.W., and W.A. Smit (2000). Failed technology futures: pitfalls and lessons from a historical survey. *Futures* **32**, 867–85.

Georgescu-Roegen, N. (1971). *The Entropy Law and the Economic Process*. Harvard University Press, Cambridge, MA.

Gowdy, J. (1994). *Coevolutionary Economics: the Economy, Society and the Environment*. Kluwer Academic Publishers, Dordrecht.

Gowdy, J. (1999). Evolution, environment and economics. In: J.C.J.M. van den Bergh (ed.), *Handbook of Environmental and Resource Economics*. Edward Elgar, Cheltenham, UK and Northampton, MA, USA, pp. 965–80.

Green, M.A. (2000). Photovoltaics: technology overview. *Energy Policy* **28**, 989–98.

Gunderson, L.H., and C.S. Holling (2002). *Panarchy: Understanding Transformations in Human and Natural Systems*. Island Press, Washington, DC.

Harmon, J. (2000). Experience curves of photovoltaic technology. Interim report IR-00-014, International Institute for Applied Systems Analysis (IIASA), Laxenburg.

Heuvel, S.T.A. van den and J.C.J.M. van den Bergh (2005). Balancing diversity and efficiency: multilevel evolutionary dynamics in the solar photovoltaic industry. Working paper, Faculty of Economics and Business Administration, Free University, Amsterdam.

Hilten, O. van, F.M.J.A. Diepstraten, D.J. Gielen and R.J. Oosterheert (1996). Energieonderzoek in Nederland. Beschrijving van inhoud, omvang en maatschappelijke aspecten, ten behoeve van de Verkenningscommissie Energieonderzoek. ECN (C-96-004), Petten.

Hirshleifer, J. (1977). Economics from a biological viewpoint. *Journal of Law and Economics* **20** (1), 1–52.

Hoang, G.T., and J. Jacquinot (2004). Controlled fusion: the next step. *Physics World*, January 2004. http://physicsweb.org/articles/world/17/1/6/1#pwhoa4_01-04

Hodgson, G.M. (1993). *Economics and Evolution: Bringing Life Back into Economics*. University of Michigan Press, Ann Arbor.

Hoed, R. van den (2004). Driving fuel cell vehicles, how established industries react to radical technologies. Thesis, Delft Technical University.

Hoeven, D. van der (2001). *Een Gedurfd Bod. Nederland zet in op de Brandstofcel*. Beta Text, Bergen.

Hoffmann, P. (2001). *Tomorrow's Energy. Hydrogen, Fuel Cells and the Prospect for a Cleaner Planet*. The MIT Press, Cambridge, MA.

Hofkes, M.W., and R. Gerlagh (2002). Escaping lock-in: the scope for a transition towards sustainable growth? Nota di Lavoro, 12.2002, Fondazione Eni Enrico Mattei, Milano, Italy.

Holland, J.H. (1998). *Emergence: from Chaos to Order*. Perseus Books, Cambridge, MA.

Huizinga, F., and P. Broer (2004). Wage moderation and labour productivity. CPB (Discussion Paper no. 28), The Hague.

IBO (2002). Samenwerken en stroomlijnen: opties voor een effectief innovatiebeleid. Eindrapportage Interdepartementaal Beleidsonderzoek (EZ-02-311), The Hague.

IEA (2002). *World Energy Outlook 2002*. International Energy Agency, Paris.

IEA (2003). Trends in photovoltaic applications. Survey report of selected IEA countries between 1992 and 2002. International Energy Agency (report IEA-PVPS T1-12: 2003), Paris.

ITER (2004). www.iter.org, accessed 13 February 2004.

Iwai, K. (1984). Schumpeterian dynamics, part I: an evolutionary model of innovation and imitation. *Journal of Economic Behaviour and Organization* **5** (2), 159–90.

Jaffe, A.B., and R.N. Stavins (1994). The energy efficiency gap: what does it mean? *Energy Policy* **22**, 804–10.

Kalhammer, F.R., P.R. Prokopius, V.P. Roan and G.E. Voecks (1998). Status and prospects of fuel cells as automobile engines: a report of the fuel cell technical advisory panel. Prepared for the State of California Air Resources Board, Sacramento, CA.

Kauffman, S.A. (1993). *The Origins of Order: Self-organization and Selection in Evolution*. Oxford University Press, Oxford.

Kemp, R. (1997). *Environmental Policy and Technical Change: a Comparison of the Technological Impact of Policy Instruments*. Edward Elgar, Cheltenham, UK.

Kemp, R., and J. Rotmans (2004). Managing the transition to sustainable mobility. In: B. Elzen, F. Geels and K. Green (eds), *System Innovation and the Transition to Sustainability: Theory, Evidence and Policy*. Edward Elgar, Cheltenham, UK and Northampton, MA, USA, pp. 137–67.

Kemp, R., J. Schot and R. Hoogma (1998). Regime shifts to sustainability through processes of niche formation: the approach of strategic niche management. *Technology Analysis & Strategic Management* **10** (2), 175–95.

Kern, S. (2000). Dutch innovation policies for the networked economy: a new approach? TNO–STB paper. TNO, Delft.

Kets, A., and G.J. Schaeffer (2004). Dutch energy R&D policy: is there a role for technology learning? In: ECN, Dutch energy policies from a European perspective, major developments in 2003. ECN (P-04-001), Petten, pp. 45–53.

Kleinknecht, A., K. van Montfort and E. Brouwer (2002). The non-trivial choice between innovation indicators. *Economic Innovations and New Technology* **11** (2), 109–21.

Knoppers, R., and G. Verbong (2001). PV in Nederland. In: G. Verbong, A. van Selm, R. Knoppers and R. Raven, *Een Kwestie van Lange Adem: de Geschiedenins van Duurzame Energie in Nederland*. Æneas, Boxtel, pp. 200–36.

Koppert, P.C., A.A. Olsthoorn and O.J. Kuik (1988). Emissiereductie door schone technologie. Een verkenning van de mogelijkheden van 10 'schone' technologieën. Report no. R-88/2, Instituut voor Milieuvraagstukken, Vrije Universiteit, Amsterdam.

KPMG (1999). Solar energy: from perennial promise to competitive alternative. KPMG Bureau voor Economische Argumentatie, Hoofddorp.

Krekel, N.R.A., P.A.M. Berdowski, and A.J. van Dieren (1987). Duurzame energie, een toekomstverkenning. Krekel van der Woerd Wouterse, Rotterdam, July 1987.

Lako, P. (ed.) (2001). Long-term scenarios and the role of fusion power. ECN (report C-01-053), Petten.

Lenstra, W.J. (2000). Zonnecellen, droom en werkelijkheid. *Economisch Statistische Berichten* **85** (4250), 288–90.

Levy, H., M. Levy and S. Solomon (2000). *Microscopic Simulation of Financial Markets: from Investor Behavior to Phenomena*. Academic Press, New York.

LSEO (1975). Interimrapport van de Landelijke Stuurgroep Energie Onderzoek. Second Chamber, The Hague (1974–75). 13250, no. 1–2.

Lumsden, C., and E.O. Wilson (1981). *Genes, Mind and Culture*. Harvard University Press, Cambridge, MA.

Lundvall, B-Å. (1992). *National Systems of Innovation*. Pinter, London.

Lyons, K.P. (1996). EU energy policies of the mid-1990s. A Business Intelligence Report. EC Inform, April 1996.

Lysen, E. (1977). *Eindeloze energie, alternatieven voor de samenleving*. Het Spectrum, Utrecht/Antwerp.

Lysen, E. (2003). Photovoltaics: an outlook for the 21st century. *Renewable Energy World*. January–February 2003, 43–53.

Maynard Smith, J. (1964). Group selection and kin selection. *Nature* **201**, 1145–6.

Maynard Smith, J. (1982). *Evolution and the Theory of Games*. Cambridge University Press, Cambridge, UK.

Maynard Smith, J., and G.R. Price (1973). The logic of animal conflict. *Nature* **246**, 15–18.

Meadows, D.H., D.L. Meadows, J. Randers and W.W. Behrens (1972). *The Limits to Growth*. Universe Books, New York.

Menkveld, M. (2004). Energietechnologieën in relatie tot transitiebeleid, factsheets. ECN (report ECN-C-04-020), Petten.

Mesner, S., and J. Gowdy (1999). Georgescu-Roegen's evolutionary economics. In: K. Mayumi and J. Gowdy (eds), *Bioeconomics and Sustainability: Essays in Honour of Nicholas Georgescu-Roegen*. Edward Elgar, Cheltenham, UK and Northampton, MA, USA, pp. 51–68.

Metcalfe, J.S. (1998). *Evolutionary Economics and Creative Destruction* (Graz Schumpeter Lectures, 1), Routledge, London.

Mokyr, J. (1990). *The Lever of the Riches: Technological Creativity and Economic Progress*. Oxford University Press, Oxford.

Mulder, P., and J.C.J.M. van den Bergh (2001). Evolutionary economic theories of sustainable development. *Growth and Change* **32** (4), 110–34.

Mulder, P., H.L.F. de Groot and M.W. Hofkes (2001). Economic growth and technological change: a comparison of insights from a neoclassical and an evolutionary perspective. *Technological Forecasting and Social Change* **68**, 151–71.

Munro, A. (1997). Economics and biological evolution. *Environmental and Resource Economics* **9**, 429–49.

Nelson, R. (ed.) (1993). *National Innovation Systems – a Comparative Analysis*. Oxford University Press, New York/Oxford.

Nelson, R., and S. Winter (1982). *An Evolutionary Theory of Economic Change*. Harvard University Press, Cambridge, MA.

Noailly, J. (2003). Coevolutionary modelling for sustainable development. Thesis, Vrije Universiteit. Thela Publishers and Tinbergen Instituut, Amsterdam.

Noailly, J., J.C.J.M. van den Bergh and C.A. Withagen (2003). Evolution of harvesting strategies: replicator and resource dynamics. *Journal of Evolutionary Economics* **13** (2), 183–200.

Norgaard, R.B. (1984). Coevolutionary development potential. *Land Economics* **60**, 160–73.

Norgaard, R.B. (1994). *Development Betrayed: the End of Progress and a Coevolutionary Revisioning of the Future*. Routledge, London/New York.

Norton, B., R. Costanza and R.C. Bishop (1998). The evolution of preferences. Why 'sovereign' preferences may not lead to sustainable policies and what to do about it. *Ecological Economics* **24**, 193–211.

NOWT (2003). Wetenschaps- en technologie-indicatoren 2003. Nederlands Observatorium van Wetenschap en Technologie (Ministry of Education, Culture and Science, MERIT, University of Leiden), The Hague.

OC&W (2003). Wetenschapsbudget 04. Second Chamber (29338, no. 1), The Hague.

Ofek, H. (2001). *Second Nature: Economic Origins of Human Evolution*. Cambridge University Press, Cambridge, UK.

Opschoor, J.B., L. de Savornin Lohman and J.B. Vos (1994). *Managing the Environment: the Role of Economic Instruments*. OECD, Paris.

Ostrom, E. (1990). *Governing the Commons: the Evolution of Institutions for Collective Action*. Cambridge University Press, Cambridge, UK.

Perlin, J. (1999), *From Space to Earth: the Story of Solar Electricity*. Aatec Publications, Ann Arbor, MI.

Poponi, D. (2003). Analysis of diffusion paths for photovoltaic technology based on experience curves. *Solar Energy* **74**, 331–40.

Porter, M. (1990). *The Competitive Advantage of Nations*. The Free Press, New York.

Porter, M., and C. van der Linde (1995). Green and competitive. *Harvard Business Review* **73**, 120–34.

Potts, J. (2000). *The New Evolutionary Microeconomics: Complexity, Competence, and Adaptive Behavior*. Edward Elgar, Cheltenham, UK and Northampton, MA, USA.

PricewaterhouseCoopers (2003). 2003 Fuel cell industry survey. A survey of 2002 financial results of North American public fuel cell companies.

Putman R.J., and S.D. Wratten (1984). *Principles of Ecology*. University of California Press, Berkeley, CA.

Quist, J. (2004). Present and past of technology and innovation policy. In: *Reader 'Technology Policy'* (Wm0914), Technical University of Delft.

Raaij, W.F. van (1988). The use of natural resources. In: W.F. van Raaij, G.M. van Veldhoven and K.E. Wärneryd (eds), *Handbook of Economic Psychology*. Kluwer Academic Publishers, Dordrecht.

Rifkin, J. (2002). *The Hydrogen Economy. The Creation of the World-Wide Energy Web and the Redistribution of Power on Earth*. Tarcher/Putnam, New York.

Riordan, M. and L. Hoddeson (1997). *Crystal Fire*. Norton, New York.

Robson, A.J. (2001). The biological basis of economic behavior. *Journal of Economic Literature* **39**, 11–33.

Roe, E.M. (1996). Sustainable development and Girardian economics. *Ecological Economics* **16**, 87–93.

Roelandt, T., P. den Hertog, J. van Sinderen and B. Vollaard (1997). Cluster analysis and cluster policy in the Netherlands. Paper presented at OECD workshop on cluster analysis and cluster policy, Amsterdam.

Rogner, H.H. (1998). Hydrogen technologies and the technology learning curve. *International Journal of Hydrogen Energy* **23** (9), 833–40.

Rotmans, J., R. Kemp and M. van Asselt (2001). More evolution than revolution: transition management in public policy, *Foresight* **3** (1), 15–31.

Rotmans, J., R. Kemp, M. van Asselt, F. Geels, G. Verbong, and K. Molendijk (2000). Transities en transitiemanagement. De casus van een emissiearme energievoorziening. ICIS/MERIT, University of Maastricht.

Samuelson, L. (1997). *Evolutionary Games and Equilibrium Selection*. The MIT Press, Cambridge, MA.

Sanders, G.A. (1972). *Energie op Leven en Dood*. Wetenschappelijke Uitgeverij, Amsterdam.

Schaeffer, G.J. (1998). Fuel cells for the future. A contribution to technology forecasting from a technology dynamics perspective. Thesis, University of Twente, Enschede.

Schlecht, L. (2003). Competition and alliances in fuel cell power train development. *International Journal of Hydrogen Energy* **28**, 717–23.

Schumpeter, J.A. (1934). *The Theory of Economic Development*. Harvard University Press, Cambridge, MA.

Schumpeter, J.A. (1939). *Business Cycles: a Theoretical, Historical and Statistical Analysis of the Capitalist Process* (2 vols). McGraw-Hill, New York.

Schumpeter, J.A. (1942). *Capitalism, Socialism and Democracy*. Harper and Brothers, New York.

Schumpeter, J.A. (1954). *Capitalism, Socialism and Democracy*, 4th edn. George Allen & Unwin, London.

SER (2003). Interactie voor innovatie. Advies 03/11. Social and Economic Council, The Hague.

Sethi, R., and E. Somanathan (1996). The evolution of social norms in common property resource use. *American Economic Review* **86** (4), 766–88.

Shell (2001). *Energy Needs, Choices and Possibilities. Scenarios to 2050*. Shell International Ltd, London.

Silverberg, G., G. Dosi and L. Orsenigo (1988). Innovation, diversity and diffusion: a self-organization model. *Economic Journal* **98**, 1032–54.

Smits, R. and S. Kuhlmann (2004). The rise of systemic instruments in innovation policy. *International Journal of Foresight and Innovation Policy* **1** (1/2), 4–32.

Solow, R.M. (1957). Technical change and the aggregate production function. *Review of Economic and Statistics* **39**, 312–20.

Somit, A. and S. Peterson (1989). *The Dynamics of Evolution: the Punctuated Equilibrium Debate in the Natural and Social Sciences.* Cornell University Press, Ithaca, NY.

Stiglitz, J.E. (1997). Reply to Daly (Forum on 'Georgescu-Roegen versus Solow/Stiglitz'). *Ecological Economics* **22**, 269–70.

Stirling, A. (2004). Diverse designs, fostering technological diversity in innovation for sustainability. Paper presented at conference 'Innovation, Sustainability and Policy', Seeon (Germany), 23–25 May 2004.

Strachan, P.A., D. Lal and F. Von Malmborg (2006). The evolving UK wind industry: critical policy and management aspects of the emerging research agenda. *European Environment* **16**, 1–18.

Strickberger, M.W. (1996). *Evolution*, 2nd edn. Jones and Bartlett, Sudbury, MA.

TNO (1975). Waterstof als energiedrager. Toekomstige mogelijkheden in Nederland. Nijverheidsorganisatie TNO, The Hague.

Tokimatsu, K., J. Fujino, S. Konishi, Y. Ogawa and K. Yamaji (2003). Role of nuclear fusion in future energy systems and the environment under future uncertainties. *Energy Policy* **31**, 775–97.

Tsuchiya, H., and O. Kobayashi (2002). Fuel cell cost study by learning curve. Paper submitted to Annual Meeting of the International Energy Workshop (EMF/IIASA), 18–20 June 2002 at Stanford University, USA.

Tullock, G. (1979). Sociobiology and economics, *Atlantic Economic Journal* **220S**, 1–10.

Tweede Kamer (2004). Nota 'De kenniseconomie in zicht': voortgangsbrief van de Minister-President. Second Chamber (27406, no. 26), The Hague.

VDEN (1980). Toekomstige Energiesituatie in Nederland. Vereniging van Directeuren van Elektriciteitsbedrijven in Nederland, Arnhem.

Veblen, T. (1898). Why is economics not an evolutionary science? *Quarterly Journal of Economics* **12** (4), 373–97.

Velthuijsen, J.W. (1995). Determinants of investment in energy conservation. Thesis, University of Groningen en SEO (Stichting voor Economisch Onderzoek), University of Amsterdam.

Voermans, F. (2004). Nedstack halveert prijs brandstofcel elk jaar; olieconcerns remmen introductie waterstof af. Utilities May 2004, pp. 12–15.

VROM (1991). Nota klimaatverandering. Tweede Kamer (22232, no. 1 and 2), The Hague.

VROM (1999). Climate policy implementation plan. Ministry of Housing, Spatial Planning and the Environment, The Hague.

VROM (2001). Where there's a will there's a world; working on sustainability. Fourth National Environmental Policy Plan (NEPP-4). Ministry of Housing, Spatial Planning and the Environment, The Hague.

VROM (2002). The progress of the Netherlands climate change policy; an assessment at the 2002 evaluation moment. Ministry of Housing, Spatial Planning and the Environment. The Hague.

VROM Council (2002). Milieu en economie: ontkoppeling door innovatie. VROM Council (advice 036), The Hague.

Weibull, J.W. (1995). *Evolutionary Game Theory*. The MIT Press, Cambridge, MA.

Westra, M.T. (2004). Kernfusie, een zon op aarde, www.fusie-energie. nl/artikelen/watisfusie.pdf, accessed 13 February 2004.

Wilson, E.O. (1998). *Consilience*. Alfred Knopf, New York.

Winder, N., B.S. McIntosh and P. Jeffrey (2005). The origin, diagnostic attributes and practical application of co-evolutionary theory. *Ecological Economics* **54** (4), 347–61.

Winter, S.G. (1964). Economic 'natural selection' and the theory of the firm. *Yale Economic Essays* **4**, 225–72.

Wit, P. de (2004). De derde generatie zonnecel uitrollen. *Shell Venster*, March/April, 26–9.

Witt, U. (ed.) (1993). *Evolutionary Economics*. The International Library of Critical Writings in Economics (Vol. 25). Edward Elgar, Cheltenham, UK and Brookfield, USA.

Zwaan, B. van der, and A. Rabl (2003). Prospects for PV: a learning curve analysis. *Solar Energy* **74**, 19–31.

Glossary

(Note that terms in italics have a separate entry.)

Adaptation   Either a state ('being adapted') or a process. As a process it denotes the adjustment through repeated selection and innovation forces acting upon a population of heterogeneous individuals or elements in an internally diverse population, with the result that over time individuals (representing strategies, behaviour, structure, technologies or products) become better adjusted to their selection environment. In evolutionary terms 'better adjusted' means a higher level of *fitness*. Adaptation as a state is the (temporary) end result of adaptation as a process. This state represents an evolutionary sequence (memory) of accumulation of innovation and selection processes in the past. See *Selection*.

Balance   See *Diversity*.

Bounded rationality   Individuals and organizations (groups) do not consistently optimize a constant and fixed exogenous utility function. Instead, they behave in accordance with adjusted or selected *routines*, imitation of others and myopia (employing a limited time horizon).

Co-evolution   A concept that originates from the synthesis of ecology and evolutionary biology. It denotes an evolutionary mechanism in which variation in one specific subsystem is subject to *selection* pressure arising from another subsystem. The concept is applied to purely biological systems (interaction between species) as well as to biological–gene–culture interactions and ecological–economic subsystem interactions. Co-evolution should be distinguished from co-dynamics; that is, non-evolutionary interaction between dynamic systems. See *Combination*.

Combination (or **recombination**)   An important source of major *innovations*. Complementary elements (technologies, organizations and institutions) form the basis and can change in joint co-operation via *co-evolution*. Cross-fertilization may be an important cause of innovative combinations, with collaboration serving as a meaningful strategy. See *Co-evolution*.

Disparity   See *Diversity*.

Diversity   A multidimensional concept describing the heterogeneity of the individuals (elements, techniques, processes, products, organizations, institutions, strategies, etc.) in a population. It covers three main dimensions; namely, *variation*, *balance* and *disparity*. Variation refers to the variety of technologies, processes, products, organizations, institutions or strategies in a population of elements. Balance (equality) denotes the distribution of the number or size of elements in a *population* – including the extent to which one or more of these elements dominate. Disparity (dissimilarity) refers to the degree of distinction between the elements in a population: are they only slightly different or very different. All three dimensions will affect *selection* outcomes as well as *innovation* potential (notably through the *combination* of current elements). See *Population*.

Fitness   A measure of the success of an individual (technology, product, strategy, etc.), in the sense of survival and extent of reproduction (including being imitated) relative to other individuals in the same population.

Increasing returns to scale   With an increase in the scale of production and sales, average profit (per unit of product sold) increases or average cost decreases. This can originate from factors on the supply or demand side of the market. Relevant factors on the supply side include economies of scale in production, learning effects, technical complementarities and network effects. On the demand side, relevant factors are learning-by-using, imitation, network effects and informational effects (the more a product has been sold, the better it is known to potential buyers). See *Lock-in* and *Path dependence*.

Innovation   The emergence of a new element (individual, technique, product, strategy, behaviour, etc.) in a population. Alternatively, it can denote the processes giving rise to it. The result is an increase in the *diversity* of elements within the population. Innovations come in various types, such as small versus gradual improvements (mutations), and major versus discrete changes (often *combinations*). See *Combination*, *Isolation* and *Serendipity*.

Innovation system   The complete set of institutional and market dimensions of innovations. This covers funding, production and use of knowledge, their various interactions, and the stages of research, development, demonstration and deployment.

Isolation   An important mechanism of *innovation*. It can be interpreted in a literal sense as spatial isolation, or in a figurative sense as cultural, social or economic isolation. *Niche markets* are an example of economic isolation.

Level playing field   A situation in which parties can compete under equal and fair conditions. From an evolutionary-economic perspective this must entail more than perfect competition. Three additional features are relevant: (i) prices must include all social – private and external – costs;

(ii) *increasing returns to scale* that favour one particular technology over another must be neutralized; (iii) promising alternatives that are still at the beginning of the learning curve must be given enough support so that they can compete seriously with technologies that are already further along on the learning curve.

Lock-in   A situation in which one product, strategy or technology dominates and from which it is impossible or very difficult to escape because of increasing returns to scale. See *Path dependence* and *Increasing returns to scale*.

Mutation   A small, gradual change that occurs at random and contributes to an innovation.

Niche market   A geographical or socio-economic market in which a certain relatively expensive (new) product or technique performs well. This may be due to spatial or socio-economic isolation (e.g., high transport costs, or the absence of a network connection, as in the case of electricity), deviating preferences or high incomes. It may also relate to a market that is protected as a result of subsidies or regulation. Niche markets assure that a new product or technology can benefit from economies of scale or learning effects. See *Isolation*.

Path dependence   A combination of preconditions, random events and *increasing returns to scale* results in a historical path that can lead to an irreversible development and possibly ultimately to the *lock-in* of a given system (technology, organizational structure, institutions, etc.). Any firm that, by coincidence, obtains a larger market share than its competitors has an advantage if this goes hand in hand with *increasing returns to scale*. This allows the firm to grow relatively quickly compared to the competitors. See *Lock-in*.

Population   A group of similar, spatially or economically connected, but nevertheless heterogeneous elements (firms, techniques, strategies, products, behaviours, etc.). These elements compete for the same resources or are active in the same market, and are subject to similar innovation mechanisms and the same *selection environment*. Evolution means that the composition of the population changes over time as a result of innovation and selection forces. As a result, evolutionary economics does not work with representative agents but with populations characterized by internal *diversity* of elements.

Recombination   See *Combination*.

Routine   The term 'organizational routines' refers to the way in which a company functions and arrives at decisions. A routine is the result of a complex synthesis of the skills of employees and their mutual relationships (communication and collaboration). Routines adapt to (economic) environments by surviving repeated selection and changing through (cumulative) intended and unintended innovations. A routine can be regarded as the equivalent of the gene in biological evolution.

Selection   Processes that reduce existing *diversity* by causing differential survival and reproduction (imitation) rates of heterogeneous elements in a population subject to selection pressure. See *Adaptation* and *Selection environment*.

Selection environment   The totality of factors that exercise selection pressure on a *population*. This includes physical, technological, geographical, business-related, market-related, institutional and regulatory factors. Economic selection factors include competition, mergers and acquisitions, reputation effects, government regulation, and financing requirements by banks. See *Selection*.

Self-organization   A process whereby interactions between elements or agents at a micro level lead to spontaneous structure and organization at a higher or macro level. Possibly, it may even result in emergence of new levels of reality. Evolution can be regarded as a particular type of self-organization.

Serendipity   An unintended discovery or invention resulting from a combination of coincidence, intelligence and knowledge (or expertise).

Transferability (transmission)   Perfect or imperfect replication occurring through reproduction or copying (imitation). These processes enable durability (retention) and cumulative change processes. In combination with selection and innovation mechanisms this can give rise to complex structures (technologies, organizations, institutions, knowledge, economic sector structure, networks and behaviours) that imply improved adaptation to a (socio-economic) environment.

Variation   See *Diversity*.

Index

ABB 106
adoption 12
Advisory Council for Science and
    Technology (AWT) 85, 96
aerospace 107, 112, 124, 133
aggregate production function 16
agriculture, co-evolution in 29, 32
Akzo Nobel 106, 112, 124
Alchian, A. 8
alkaline (AFC) fuel cells 104,
    106
alternative engine fuels 77
applied research 23, 84
arms race 31
artificial life 14
Astro Solar 124
asymptotic equilibria 13
automatic behaviour 11
Automatic Power 134–5
autonomous evolution 49
autonomous transitions 38

backcasting 91
balance 21, 143, 145
    see also diversity
balance of System 132
Ballard 107, 109, 112
behaviour, transfer of 8
behavioural change 9, 51
behavioural economics 27–8
biomass fuel 86
BioPartner 84
blueprints 10
bounded rationality 1, 3, 11, 18, 27–8,
    33–4, 53–6, 93–5, 113–14, 120–21,
    135, 141, 142, 147
    see also Girardian economics;
        imitation; myopia; routines
BP 124
BSIK fund 84
Bureau of Industrial Properties
    (Netherlands) 81

business community, nuclear fusion
    development 118
business cycles 7
business routines 14
Buying green, a handbook on
    environmental public procurement
    (EU document) 93

carbon dioxide *see* CO_2 capture; CO_2
    emissions
cars, fuel cell technology 106–7, 108–9
catch-up mechanism 17
CENELEC 129
centralized energy supply 50
chain efflciency, sustainable energy 77
chance events 34
changes, in routine 10–11
Chicago economists 13
climate policy, Netherlands 70, 73, 74
cluster policy, technology development
    79
CO_2 capture 55
CO_2 emissions 55, 73, 74, 93, 104, 119
co-adaptation 30
co-evolution 29–32, 36, 47, 62, 98–9,
    114, 121, 135, 141–2
co-financing 89
co-operation 89–90, 112, 122, 137, 150
Coase negotiation theorem 41
codes, PV market 129
combination 22–3, 34, 87, 91, 112, 133,
    144, 150
    see also recombination
combinatorial mathematics 14
combined heat and power (CHP) 55,
    143
commercial nuclear fusion reactors 119
common property, overexploitation of
    44
companies
    innovations and asymmetry between
        12